Swift High Performance

Leverage Swift and enhance your code to take your applications to the next level

Kostiantyn Koval

PUBLISHING

BIRMINGHAM - MUMBAI

Swift High Performance

First published: November 2015

Production reference: 1271015

Published by Packt Publishing Ltd.
Livery Place
35 Livery Street
Birmingham B3 2PB, UK.

ISBN 978-1-78528-220-1

www.packtpub.com

Credits

Author
Kostiantyn Koval

Reviewers
Ravi Shankar

Tatsuya Tobioka

Ye Xiaodong

Acquisition Editor
Reshma Raman

Content Development Editor
Dharmesh Parmar

Technical Editor
Shiny Poojary

Copy Editors
Stephen Copestake

Vikrant Phadke

Project Coordinator
Izzat Contractor

Proofreader
Safis Editing

Indexer
Rekha Nair

Production Coordinator
Melwyn Dsa

Cover Work
Melwyn Dsa

About the Author

Kostiantyn Koval is a passionate developer with 5 years of experience. All the time, his main passion and work has been building iOS applications. So far, he has built many different applications, including games, enterprise apps, and big platforms. He fell in love with Swift the first minute he saw it, and keeps expressing this to the rest of the world.

Other than iOS, he is also interested in technologies and languages such as Scala, Clojure, LLVM, Ruby, JavaScript, and others.

He loves open source and blogging. You can find him on GitHub at `https://github.com/kostiakoval` and his blogs at `https://medium.com/@kostiakoval`. Other places to contact him are `http://kostiakoval.github.io` and Twitter at `@KostiaKoval`.

His hobbies are programming, building start-ups, and making this world better with software.

I really want to say thanks to my wife, Tetiana, for believing in me, supporting me, and helping me in life, especially during the writing of this book.

About the Reviewers

Ravi Shankar is a multi-skilled software consultant with over 15 years of experience in the IT industry. He has good all-around ability in different technologies and extensive experience in product development, system maintenance, and support. He is a polyglot and self-taught programmer with hands-on experience in Swift, Objective-C, and Java. Ravi believes in gaining knowledge through sharing and helping others learn.

Thanks to Izzat and Packt Publishing for giving me this opportunity.

Tatsuya Tobioka is a software engineer familiar with Ruby, JavaScript, Objective-C, and Swift. He lives happily with his beloved wife and children in Tokyo, Japan.

He started iOS development in 2010, and then released a number of apps for developers, such as JavaScript Anywhere, Edhita, and CoffeeScript At Once.

Currently, Tatsuya spends much of his time learning tvOS.

You can check out his open source projects on GitHub at `https://github.com/tnantoka` and tweets at `@tnantoka`.

Ye Xiaodong is a full-stack software engineer and technical director of zai360.com (http://www.zai360.com/), an O2O company that provides recyclable collection services for Chinese family customers on a periodical basis. He has 7 years of application development experience working for start-ups and leading companies across the world with iOS, Android, Windows Phone, Symbian, and Meego. He has developed lots of iOS applications; designed, created, and maintained iOS libraries and Xcode plugins; and contributed to open source projects. Ye is passionate about bringing the latest features into applications. He was a technical reviewer for *Mastering Swift and Swift Design Patterns, Packt Publishing*.

First and foremost, I would like to thank the coordinator of this project, Izzat Contractor, for her valuable guidance and advice.

www.PacktPub.com

Support files, eBooks, discount offers, and more

For support files and downloads related to your book, please visit www.PacktPub.com.

Did you know that Packt offers eBook versions of every book published, with PDF and ePub files available? You can upgrade to the eBook version at www.PacktPub.com and as a print book customer, you are entitled to a discount on the eBook copy. Get in touch with us at service@packtpub.com for more details.

At www.PacktPub.com, you can also read a collection of free technical articles, sign up for a range of free newsletters and receive exclusive discounts and offers on Packt books and eBooks.

https://www2.packtpub.com/books/subscription/packtlib

Do you need instant solutions to your IT questions? PacktLib is Packt's online digital book library. Here, you can search, access, and read Packt's entire library of books.

Why subscribe?

- Fully searchable across every book published by Packt
- Copy and paste, print, and bookmark content
- On demand and accessible via a web browser

Free access for Packt account holders

If you have an account with Packt at www.PacktPub.com, you can use this to access PacktLib today and view 9 entirely free books. Simply use your login credentials for immediate access.

Table of Contents

Preface

During the WWDC in June 2014, Apple announced a new programming language, called Swift. Swift is a very modern and powerful language. In the last year, Swift has become a very popular programming language. It has evolved and changed. Since Swift is fairly new, there are many questions related to its performance characteristics and best practice for achieving high performance in it.

Swift High Performance provides an overview of the important features of Swift, its performance characteristics, and sets of advices and techniques that allow you to build solid and sustainable applications in Swift with great performance.

This book also provides an overview of different tools that help you debug, investigate, and improve your code.

What this book covers

Chapter 1, Exploring Swift's Power and Performance, introduces Swift, its powerful features, its performance, and its interoperability with Objective-C.

Chapter 2, Making a Good Application Architecture in Swift, covers in detail the powerful features of Swift and how to apply them to build a solid application architecture.

Chapter 3, Testing and Identifying Slow Code with the Swift Toolkit, introduces different Swift and Xcode tools for code prototyping, performance measurement, and identifying and improving slow code.

Chapter 4, Improving Code Performance, shows Swift's performance-related details and features and demonstrates how Swift achieves its high performance. This chapter also covers different optimization techniques for improving performance in your applications.

Chapter 5, *Choosing the Correct Data Structure*, covers different data structures, their features, their performance characteristics, and suggestions on when to apply them.

Chapter 6, *Architecting Applications for High Performance*, demonstrates different application architecture techniques that allow you to achieve high performance, such as concurrency, avoiding state, and single responsibility.

Chapter 7, *The Importance of Being Lazy*, covers important techniques for improving an application's performance, such as lazy loading, lazy collections, and evaluation.

Chapter 8, *Discovering All the Underlying Swift Power*, gives you more details of Swift's structure, its tools, and the compilation process, and gives a better understanding of how Swift achieves its performance.

What you need for this book

This book's content and code examples were written using Xcode 7 and Swift 2.0. To follow along with the tutorials, you will need the following:

- **Mac OS 10.9 or later**: Currently the Swift IDE, compiler, and tools are only available for Mac OS.

- **Xcode 7.0 or later**: Xcode is the main development tool for Swift iOS and Mac apps. It can be installed via the Mac AppStore at `https://itunes.apple.com/en/app/xcode/id497799835?mt=12`.

- **Command-line tools for Xcode and simulators**: Once you have installed Xcode and launched it, it will offer to install additional command-line tools. Xcode installs simulators by default, but you can download more simulators by going to **Xcode** | **Preferences** | **Download**.

Who this book is for

This book is for developers who already know the basics of Swift and want to learn the more advanced features and tips on how to achieve high performance and build solid applications in Swift. We assume that you are familiar with at least a bit of Mac OS and the Xcode IDE. This book is for everyone who wants to takes their knowledge of Swift to a new level.

Knowledge of iOS or Mac OS programming and Objective-C would be plus, but it's not required.

Conventions

In this book, you will find a number of text styles that distinguish between different kinds of information. Here are some examples of these styles and an explanation of their meaning.

Code words in text, database table names, folder names, filenames, file extensions, pathnames, dummy URLs, user input, and Twitter handles are shown as follows: "First, let's add a nickname to the `Person` class."

A block of code is set as follows:

```
var sam = Person(firstName: "Sam", lastName: "Bosh",
  nickName:"BigSam")
sam = sam.changeNickName("Rockky")
```

When we wish to draw your attention to a particular part of a code block, the relevant lines or items are set in bold:

```
<type parameter : constraint >

func minElem<T : Comparable>(x: T, _ y: T) -> T {
  return x < y ? x : y
}
```

Any command-line input or output is written as follows:

```
(lldb) repl
1> func isAllPositive(ar: [Int]) -> Bool {
2.    let negatives = ar.filter { $0 < 0 }
3.    return negatives.count == 0
4. }
```

New terms and **important words** are shown in bold. Words that you see on the screen, for example, in menus or dialog boxes, appear in the text like this: "We will choose a **Time Profiler** template and click on **Record**."

 Warnings or important notes appear in a box like this.

 Tips and tricks appear like this.

Reader feedback

Feedback from our readers is always welcome. Let us know what you think about this book—what you liked or disliked. Reader feedback is important for us as it helps us develop titles that you will really get the most out of.

To send us general feedback, simply e-mail `feedback@packtpub.com`, and mention the book's title in the subject of your message.

If there is a topic that you have expertise in and you are interested in either writing or contributing to a book, see our author guide at `www.packtpub.com/authors`.

Customer support

Now that you are the proud owner of a Packt book, we have a number of things to help you to get the most from your purchase.

Downloading the example code

You can download the example code files from your account at `http://www.packtpub.com` for all the Packt Publishing books you have purchased. If you purchased this book elsewhere, you can visit `http://www.packtpub.com/support` and register to have the files e-mailed directly to you, or you can visit `https://github.com/kostiakoval/SwiftHighPerformance`, which is the GitHub repo with code examples for this book.

Errata

Although we have taken every care to ensure the accuracy of our content, mistakes do happen. If you find a mistake in one of our books—maybe a mistake in the text or the code—we would be grateful if you could report this to us. By doing so, you can save other readers from frustration and help us improve subsequent versions of this book. If you find any errata, please report them by visiting `http://www.packtpub.com/submit-errata`, selecting your book, clicking on the **Errata Submission Form** link, and entering the details of your errata. Once your errata are verified, your submission will be accepted and the errata will be uploaded to our website or added to any list of existing errata under the Errata section of that title.

To view the previously submitted errata, go to `https://www.packtpub.com/books/content/support` and enter the name of the book in the search field. The required information will appear under the **Errata** section.

Piracy

Piracy of copyrighted material on the Internet is an ongoing problem across all media. At Packt, we take the protection of our copyright and licenses very seriously. If you come across any illegal copies of our works in any form on the Internet, please provide us with the location address or website name immediately so that we can pursue a remedy.

Please contact us at copyright@packtpub.com with a link to the suspected pirated material.

We appreciate your help in protecting our authors and our ability to bring you valuable content.

Questions

If you have a problem with any aspect of this book, you can contact us at questions@packtpub.com, and we will do our best to address the problem.

1

Exploring Swift's Power and Performance

In the 2014, Apple released a new programming language, called Swift. Swift has been designed from scratch with many powerful features. It is statically typed and very safe. It has a clean and nice syntax, it's fast, it's flexible, and it has many other advantages that you will learn later in the book. Swift seems to be very powerful and it has big potential. Apple has set big expectations for Swift, and their main goal for Swift is that it should be a replacement for Objective-C, which is going to happen in the near future.

In this chapter, you will become familiar with the Swift programming language, what it was made for, and what its advantages and features are. We will also make our first Swift application and see how easy it is to integrate with existing Objective-C code.

In this chapter, we will cover the following topics:

- Welcome to Swift
- Writing swift code
- Swift interoperability
- The importance of performance and performance key metrics

Swift speed

I can guess you opened this book because you are interested in speed and are probably wondering, "How fast can Swift be?" Before you even start learning Swift and discovering all the good things about it, let's answer it right here and right now.

Let's take an array of 100,000 random numbers; sort it in Swift, Objective-C, and C using the standard `sort` function from `stdlib` (`sort` in Swift, `qsort` in C, and `compare` in Objective-C); and measure how much time each would take.

Sorting an array with 100,000 integer elements gives us this:

Technology	Time taken
Swift	0.00600 sec
C	0.01396 sec
Objective-C	0.08705 sec

And the winner is, Swift! Swift is **14.5** times faster than Objective-C and **2.3** times faster than C.

In other examples and experiments, C is usually faster than Swift and Swift is way faster than Objective-C. These measurements were done with Xcode 7.0 beta 6 and Swift 2.0. It's important to highlight that the improvements in Swift 2.0 were mainly focused on making it cleaner, more powerful, safer, and more stable, and preparing it for open sourcing. Swift's performance hasn't reached its full potential yet, and the future is so exciting!

Welcome to Swift

The Swift programming language has been designed by Apple from the ground up. It was released with the slogan *Objective-C without the C*. The meaning of this phrase is that Swift doesn't have any limitation of backward compatibilities. It's totally new and with no old baggage. Before you start learning all the power of Swift, I think it would be useful to answer a few questions about why should you learn it, and if you have any doubts about that, I should dispel them.

Why should I learn Swift?

Swift is a very new programming language but it has become very popular and has gained huge traction. However, many iOS and OS X developers ask these questions:

- Should I learn Swift?
- What should I learn, Swift or Objective-C?
- Is Objective-C going to stay or die?
- Is Swift ready for production apps?
- Is Swift faster than Objective-C or C?
- What applications can I write using Swift?

My answer is, "Yes. Definitely!" You should learn Swift. It doesn't matter whether you are a new iOS and OS X developer or you have some Objective-C background; you should definitely learn Swift.

If you are new developer, then it's really useful to start with Swift, because you will learn programming basics and techniques in Swift, and further Swift learning would be much easier. Although it would definitely useful to learn Objective-C as well, I would recommend learning Swift first so that you build your programming mindset on Swift.

If you already have some experience in Objective-C, then you should try Swift as soon as possible. It will not only give you the knowledge of a new programming language, but also open the door to new ideas and ways of solving problems in Objective-C. We can see that Objective-C has started evolving right now because of Swift.

Objective-C has many limitations because of its backward capabilities with C. It was created 23 years ago, in 1983, but it will die much sooner than Swift.

After the release of Swift version 1.0, in only a year's time we have seen many Swift applications successfully developed and released on the App Store. In this time period, many Swift tools and open source libraries that increase development productivity have been created.

During WWDC 2015, Apple announced that Swift will be made open source. This means that Swift can be used to write any software and not only iOS or OS X apps. You can write a piece of server-side code or web app in Swift. This is one more reason you should learn it.

On the other hand, we see that Swift is under constant development. There were many changes and improvements in version 1.2, and there were even more changes in version 2.0. Although it's very easy to upgrade to the newer Swift version with the Xcode migrator, it's something you should think about.

Swift has some promising performance characteristics. We have seen a huge performance improvement in the Swift 1.2 release, and some improvements in Swift 2.0 as well. You have seen from the previous example how fast Swift is, and in general, Swift has more potential to achieve high performance than Objective-C.

Finally, I want to mention a phrase I really like, by Bryan Irace:

> *When the iOS SDK says "Jump", ask "How High?"*

Don't wait, learn Swift!

Swift's features and benefits

At this point, you know that you should learn Swift, and you shouldn't have any doubts. Let's take a look what makes Swift so amazing and powerful. Here is a list of a few important features that we are going to cover:

- Clean and beautiful syntax
- Type-safe
- Reach types system
- Powerful value types
- A multiparadigm language—object-oriented, protocol-oriented, and functional
- Generic purpose
- Fast
- Safe

Clean and beautiful

Powerful features and performance are important, but I think that cleanness and beauty are no less important. You write and read code everyday, and it has to be clean and beautiful so that you can enjoy it. Swift is very clean and beautiful, and the following are the main features that make it so.

No semicolons

Semicolons were created for the compiler. They help the compiler understand the source code and split it into commands and instructions. But the source code is written for people, and we should probably get rid of the compiler instructions from it:

```
var number = 10
number + 5

// Not recommended
var count = 1;
var age = 18; age++
```

There is no need for a semicolon (;) at the end of every instruction. It may seem like a very small feature, but it makes code so much nicer and easier to write and read. You can, however, put semicolons if you want. A semicolon is required when you have two instructions on the same line. There are also some exceptions when you have to use semicolons, a `for` loop as an example (`for var i = 0; i < 10; i++`), but in that context, they are used for a different purpose.

 I strongly recommend not using semicolons, and avoid using more than one instruction in the same line.

Type inference

With type inference, you don't need to specify the types of variables and constants. Swift automatically detects the correct type from the context. Sometimes, however, you have to specify the type explicitly and provide type annotation. When there is no value assigned to the variable, Swift can't predict what type that variable should be:

```
var count = 10          //count: Int
var name = "Sara"       //name: String
var empty = name.isEmpty   //empty: Bool

// Not recommended
var count: Int = 10
var name: String = "Sara"
var empty: Bool = name.isEmpty

// When you must provide type annotation
var count: Int
var name: String

count = 10
name = "Sara"
```

In most cases, Swift can understand a variable's type from the value assigned to it.

 Don't use type annotation if it's not required. Giving your variables descriptive names should be enough. This makes your code clean and nice to read.

Other clean code Swift features

The list of all of Swift's clean code features is very long; here are few of them: closure syntax, functions' default parameter values, functions' external parameter names, default initializers, subscripts, and operators:

- **Clean closure syntax**: A closure is a standalone block of code that can be treated as a light unnamed function. It has the same functionality as a function but has a cleaner syntax. You can assign it to a variable, call it, or pass it as an argument to a function. For example, `{ $0 + 10 }` is a closure:

```
let add10 = { $0 + 10 }
add10(5)

let numbers = [1, 2, 3, 4]
numbers.map { $0 + 10 }
numbers.map(add10)
```

- **Default parameter values and external names**: While declaring a function, you can define default values for parameters and give them different external names, which are used when you call that function. With default parameters, you can define one function but call it with different arguments. This reduces the need for creating unnecessary functions:

```
func complexFunc (x: Int, _ y: Int = 0,
  extraNumber z: Int = 0, name: String = "default") -> String{
    return  "\(name): \(x) + \(y) + \(z) = \(x + y + z)"
}

complexFunc(10)
complexFunc(10, 11)
complexFunc(10, 11, extraNumber: 20, name: "name")
```

- **Default and memberwise initializers**: Swift can create initializers for struct and base classes in some scenarios for you. Less code, better code:

```
struct Person {
    let name: String
    let lastName: String
    let age: Int
}


Person(name: "Jon", lastName: "Bosh", age: 23)
```

- **Subscripts**: This is a nice way of accessing the member elements of a collection. You can use any type as a key:

```
let numbers = [1, 2, 3, 4]
let num2 = numbers[2]

let population = [
  "China" : 1_370_940_000,
  "Australia" : 23_830_900
]
population["Australia"]
```

You can also define a subscript operator for your own types or extend existing types by adding own subscript operator to them in an extension:

```
// Custom subscript
struct Stack {
  private var items: [Int]

  subscript (index: Int) -> Int {
    return items[index]
  }

  // Stack standard functions
  mutating func push(item: Int) {
    items.append(item)
  }

  mutating func pop() -> Int {
    return items.removeLast()
  }
}

var stack = Stack(items: [10, 2])
stack.push(6)
stack[2]
stack.pop()
```

- **Operators**: These are symbols that represent functionality, for example, the + operator. You can extend your types to support standard operators or create your own custom operators:

```
let numbers = [10, 20]
let array = [1, 2, 3]
let res = array + numbers

struct Vector {
   let x: Int
   let y: Int
}

func + (lhs: Vector, rhs: Vector) -> Vector {
   return Vector(x: lhs.x + rhs.x, y: lhs.y + rhs.y);
}

let a = Vector(x: 10, y: 5)
let b = Vector(x: 2, y: 3)

let c = a + b
```

 Define your custom operators carefully. They can make code cleaner, but they can also bring much more complexity into the code and make it hard to understand.

- **guard**: The guard statement is used to check whether a condition is met before continuing to execute the code. If the condition isn't met, it must exit the scope. The guard statement removes nested conditional statements and the Pyramid of Doom problem:

 Read more about the Pyramid of Doom at https://en.wikipedia. org/wiki/Pyramid_of_doom_(programming).

```
func doItGuard(x: Int?, y: Int) {
   guard let x = x else { return }
   //handle x
   print(x)

   guard y > 10 else { return }
   //handle y
   print(y)
}
```

A clean code summary

As you can see, Swift is very clean and nice. The best way to show how clean and beautiful Swift is is by trying to implement the same functionality in Swift and Objective-C.

Let's say we have a list of people and we need to find the people with a certain age criteria and make their names lowercase.

This is what the Swift version of this code will look like:

```swift
struct Person {
  let name: String
  let age: Int
}

let people = [
  Person(name: "Sam", age: 10),
  Person(name: "Sara", age: 24),
  Person(name: "Ola", age: 42),
  Person(name: "Jon", age: 19)
]

let kids = people.filter { person in person.age < 18 }
let names = people.map { $0.name.lowercaseString }
```

The following is what the Objective-C version of this code will look like:

```objc
//Person.h File
@import Foundation;

@interface Person : NSObject

@property (nonatomic) NSString *name;
@property (nonatomic) NSInteger age;

- (instancetype)initWithName:(NSString *)name age:(NSInteger)age;

@end

//Person.m File
#import "Person.h"

@implementation Person

- (instancetype)initWithName:(NSString *)name age:(NSInteger)age {
```

```
    self = [super init];
    if (!self) return nil;

    _name = name;
    _age = age;

    return self;
}

@end

NSArray *people = @[
    [[Person alloc] initWithName:@"Sam" age:10],
    [[Person alloc] initWithName:@"Sara" age:24],
    [[Person alloc] initWithName:@"Ola" age:42],
    [[Person alloc] initWithName:@"Jon" age:19]
];

NSArray *kids = [people filteredArrayUsingPredicate:[NSPredicate
  predicateWithFormat:@"age < 18"]];

NSMutableArray *names = [NSMutableArray new];
for (Person *person in people) {
  [names addObject:person.name.lowercaseString];
}
```

The results are quite astonishing. The Swift code has 14 lines, whereas the Objective-C code has 40 lines, with `.h` and `.m` files. Now you see the difference.

Safe

Swift is a very safe programming language, and it does a lot of security checks at compile time. The goal is to catch as many issues as possible during compiling and not when you run an application.

Swift is a type-safe programming language. If you made any mistakes with a type, such as trying to add an `Int` and a `String` or passing the wrong argument to a function, you will get an error:

```
let number = 10
let part = 1.5

number + part; // Error

let result = Double(number) + part
```

Swift doesn't do any typecasting for you; you have to do it explicitly, and this makes Swift even safer. In this example, we had to cast an `Int` number to the `Double` type before adding it.

Optionals

A very important safe type that was introduced in Swift is an **optional**. An optional is a way of representing the absence of a value—`nil`. You can't assign `nil` to a variable with the `String` type. Instead, you must declare that this variable can be `nil` by making it the optional `String?` type:

```
var name: String = "Sara"
name = nil //Error. You can't assign nil to a non-optional type

var maybeName: String?
maybeName = "Sara"
maybeName = nil // This is allowed now
```

To make a type an optional type, you must put a question mark (?) after the type, for example, `Int?`, `String?`, and `Person?`.

You can also declare an optional type using the `Optional` keyword, `Optional<String>`, but the shorter way with using `?` is preferred:

```
var someName: Optional<String>
```

Optionals are like a box that contains some value or nothing. Before using the value, you need to unwrap it first. This technique is called unwrapping optionals, or optional binding if you assign an unwrapped value to a constant:

```
if let name = maybeName {
  var res = "Name - " + name
} else {
  print("No name")
}
```

 You must always check whether an optional has a value before accessing it.

Error handling

Swift 2.0 has powerful and very simple-to-use error handling. Its syntax is very similar to the exception handling syntax in other languages, but it works in a different way. It has the `throw`, `catch`, and `try` keywords. Swift error handling consists of a few components, explained as follows:

- An error object represents an error, and it must conform to the `ErrorType` protocol:

```
enum MyErrors: ErrorType {
  case NotFound
  case BadInstruction
}
```

 Swift enumerations fit best for representing a group of related error objects.

- Every function that can throw an error must be declared using the `throws` keyword after its parameters' list:

```
func dangerous(x: Int) throws
func dangerousIncrease(x: Int) throws -> Int
```

- To throw an error, use the `throw` keyword:

```
throw MyErrors.BadInstruction
```

- When you are calling a function that can throw an error, you must use the `try` keyword. This indicates that a function can fail and further code will not be executed:

```
try dangerous(10)
```

- If an error occurs, it must be caught and handled with the `do` and `try` keywords or thrown further by declaring that function with `throws`:

```
do {
  try dangerous(10)
}
catch {
  print("error")
}
```

Let's take a look at a code example that shows how to work with exceptions in Swift:

```swift
enum Error: ErrorType {
  case NotNumber(String)
  case Empty
}

func increase(x: String) throws -> String {
  if x.isEmpty {
    throw Error.Empty
  }

  guard let num = Int(x) else {
    throw Error.NotNumber(x)
  }

  return String(num + 1)
}

do {
  try increase("10")
  try increase("Hi")
}
catch Error.Empty {
  print("Empty")
}
catch Error.NotNumber (let string) {
  print("\"\(string)\" is not a number")
}
catch {
  print(error)
}
```

There are many other safety features in Swift:

- Memory safety ensures that values are initialized before use.
- Two-phase initialization process with security checks
- Required method overriding and many others

Rich type system

Swift has the following powerful types:

- **Structures** are flexible building blocks that can hold data and methods to manipulate that data. Structures are very similar to classes but they are value type:

```
struct Person {
    let name: String
    let lastName: String

    func fullName() -> String {
        return name + " " + lastName
    }
}

let sara = Person(name: "Sara", lastName: "Johan")
sara.fullName()
```

- **Tuples** are a way of grouping multiple values into one type. Values inside a tuple can have different types. Tuples are very useful for returning multiple values from a function. You can access values inside a tuple by either index or name if the tuple has named elements; or you can assign each item in the tuple to a constant or a variable:

```
let numbers = (1, 5.5)
numbers.0
numbers.1

let result: (code: Int, message: String) = (404, "Not fount")
result.code
result.message

let (code ,message) = (404, "Not fount")
```

- **Range** represents a range of numbers from x to y. There are also two range operators that help create ranges: closed range operator and half-open range operator:

```
let range = Range(start: 0, end: 100)
let ten = 1...10 //Closed range, include last value 10
let nine = 0..<10 //half-open, not include 10
```

- **Enumeration** represents a group of common related values. An enumeration's member can be empty, have a raw value, or have an associated value of any type. Enumerations are first-class types; they can have methods, computed properties, initializer, and other features. They are great for type-safe coding:

```
enum Action: String {
   case TakePhoto
   case SendEmail
   case Delete
}

let sendEmail = Action.SendEmail
sendEmail.rawValue //"SendEmail"

let delete = Action(rawValue: "Delete")
```

Powerful value types

There are two very powerful value types in Swift: `struct` and `enum`. Almost all types in the Swift standard library are implemented as immutable value types using `struct` or `enum`, for example, `Range`, `String`, `Array`, `Int`, `Dictionary`, `Optionals`, and others.

Value types have four big advantages over reference types, they are:

- Immutable
- Thread safe
- Single owned
- Allocated on the stack memory

Value types are immutable and only have a single owner. The value data is copied on assignment and when passing it as an argument to a function:

```
var str = "Hello"
var str2 = str

str += " :)"
```

 Swift is smart enough to perform value copying only if the value is mutated. Value copying doesn't happen on an assignment, that is `str2 = str`, but on value mutation, that is `str += ":)"`. If you remove that line of code, `str` and `str2` would share the same immutable data.

A multiparadigm language

Swift is a multiparadigm programming language. It supports many different programming styles, such as object-oriented, protocol-oriented, functional, generic, block-structured, imperative, and declarative programming. Let's take a look at a few of them in more detail here.

Object oriented

Swift supports the object-oriented programming style. It has classes with the single inheritance model, the ability to conform to protocols, access control, nested types and initializers, properties with observers, and other features of OOP.

Protocol oriented

The concept of protocols and protocol-oriented programming is not new, but Swift protocols have some powerful features that make them special. The general idea of protocol-oriented programming is to use protocols instead of types. In this way, we can create a very flexible system with weak binding to concrete types.

In Swift, you can extend protocols and provide a method's default implementation:

```
extension CollectionType {

  func findFirst (find: (Self.Generator.Element) -> Bool) ->
    Self.Generator.Element? {

    for x in self {
      if find(x) {
        return x
      }
    }
    return nil
  }
}
```

Now, every type that implements `CollectionType` has a `findFirst` method:

```
let a = [1, 200, 400]
let r = a.findFirst { $0  > 100 }
```

One big advantage of using protocol-oriented programming is that we can add methods to related types and use the dot (.) syntax for method chaining instead of using free functions and passing arguments:

```
let ar = [1, 200, 400]

//Old way
map(filter(map(ar) { $0 * 2 }) { $0 > 50 }) { $0 + 10 }

//New way
ar.map{ $0 * 2 } .filter{ $0 > 50 } .map{ $0 + 10 }
```

Functional

Swift also supports the functional programming style. In functional languages, a function is a type and it is treated in the same way as other types, such as Int; also, it is called a **first class function**. Functions can be assigned to a variable and passed as an argument to other functions. This really helps to decouple your code and makes it more reusable.

A great example is a filter function of an array. It takes a function that performs the actual filtering logic, and it gives us so much flexibility:

```
// Array filter function from Swift standard library
func filter(includeElement: (T) -> Bool) -> [T]

let numbers = [1, 2, 4]

func isEven (x: Int) -> Bool {
    return x % 2 == 0
}
let res = numbers.filter(isEven)
```

Generic purpose

Swift has a very powerful feature called **generics**. Generics allow you to write generic code without mentioning a specific type that it should work with. Generics are very useful for building algorithms, reusable code, and frameworks. The best way to explain generics is by showing an example. Let's create a minimum function that will return a smaller value:

```
func minimum(x: Int, _ y: Int) -> Int {
    return (x < y) ? x : y
}

minimum(10, 11)
minimum(11,5, 14.3) // error
```

This function has a limitation; it will work only with integers. However, the logic of getting a smaller value is the same for all types—compare them and return the smaller value. This is very generic code.

Let's make our `minimum` function generic and work with different types:

```
func minimum <T : Comparable>(x: T, _ y: T) -> T {
  return (x < y) ? x : y
}

minimum (10, 11)
minimum (10.5, 1.4)
minimum ("A", "ABC")
```

[The Swift standard library has already implemented a generic `min` function. Use that instead.]

Fast

Swift is designed to be fast and have high performance, and this is achieved with the following techniques:

- Compile-time method binding
- Strong typing and compile time optimization
- Memory layout optimization

Later, we will cover in more detail how Swift uses these techniques to improve performance.

Swift interoperability

There are two main points that Apple thought of when introducing Swift:

- The usage of the Cocoa framework and established Cocoa patterns
- Easy to adopt and migrate

Apple understood that and took it very seriously while working on Swift. They made Swift work seamlessly with Objective-C and Cocoa. You can use all Objective-C code in Swift, and you can even use Swift in Objective-C.

It's very crucial to be able to use the Cocoa framework. All of the code that is written in Objective-C is available for use in Swift, both Apple frameworks and third-party libraries as well.

Using Objective-C in Swift

All the Cocoa frameworks written in Objective-C are available in Swift by default. You just need to import them and then use them. Swift doesn't have header files; instead, you need to use a module name. You can also include your own Swift frameworks in the same way:

```
import Foundation
import UIKit
import Alamofire // Custom framework
```

Setup

To include your own Objective-C source files, you need to do a small setup first. The process is a bit different for the application target and framework target. The main idea is the same — to import the Objective-C header files.

The application target

For the application target, you need to create a bridging header. A bridging header is a plain Objective-C header file in which you specify the Objective-C `import` statements.

Xcode will show a popup, offering to create, and set up a bridging header for you when you add the Objective-C file to a Swift project, or vice versa for the first time. This is the best and the most convenient way to add it.

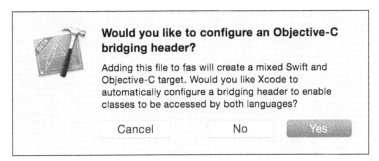

Would you like to configure an Objective-C bridging header?

Adding this file to fas will create a mixed Swift and Objective-C target. Would you like Xcode to automatically configure a bridging header to enable classes to be accessed by both languages?

Cancel   No   Yes

If you decline the Xcode help, you can create a bridging header yourself anytime. To do that, you need to follow these steps:

1. Add a new header file to the project.
2. Go to **Target | Build Settings**.
3. Search for `Objective-C Bridging Header` and specify the path to the bridging header file created in step 1.

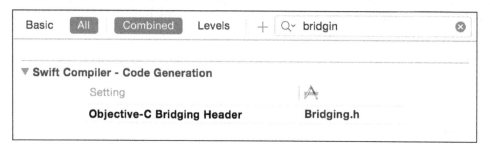

Once you set up bridging header, the next step is to add `import` statements to it:

`Bridging.h`

```
//
//  Use this file to import your target's public headers that you
//  would like to expose to Swift.

#import "MyClass.h"
```

The framework target

For the framework target, you simply need to import the `.h` Objective-C header files to the framework's umbrella header. The Objective-C header files must be marked as public. The umbrella header is the header in which you specify your publicly available API. Usually, it looks like this—the `ExampleFramework.h` umbrella header:

```
#import <UIKit/UIKit.h>

//! Project version number for MySwiftKit.
FOUNDATION_EXPORT double MySwiftKitVersionNumber;

//! Project version string for MySwiftKit.
FOUNDATION_EXPORT const unsigned char MySwiftKitVersionString[];

// In this header, you should import all the public headers of your
framework using statements like #import <MySwiftKit/PublicHeader.h>

#import <SimpleFramework/MyClass.h>
```

Calling Objective-C code

Once you are done with the setup, you can use all Objective-C APIs in Swift. You can create instances, call methods, inherit from Objective-C classes, conform to protocols, and do other things that you can do in Objective-C. In this example, we will use the Foundation classes, but the rules are the same for third-party code as well:

```
import UIKit
import Foundation

let date = NSDate()
date.timeIntervalSinceNow

UIColor.blackColor()
UIColor(red: 0.5, green: 1, blue: 1, alpha: 1)

class MyView: UIView {
    //custom implementation
}
```

 Inherit from Objective-C classes only if you need it. This can have a negative impact on performance.

There is free bridging between Swift types and Objective-C Foundation types. Automatic bridging happens on assignment and when you pass it as an argument to a function:

```
let array = [1, 2, 3]

func takeArray(array: NSArray) { }

var objcArray: NSArray = array
takeArray(array)
```

Converting from Objective-C to a Swift type requires explicit type casting. There are two types of casting: downcasting and upcasting. Casting is usually an unsafe operation, which could fail, and that's why it returns an optional type:

```
//Upcasting or safe casting
let otherArray: [AnyObject] = objcArray as [AnyObject]

//Downcasting, unsafe casting
if let safeNums = objcArray as? [Int] {
  safeNums[0] + 10 //11
}

let string: NSString = "Hi"
let str: String = string as String
```

The `String` type has gone one step even further. You can invoke the Objective-C foundation methods on the Swift `String` type without any type casting:

```
var name: String = "Name"
name.stringByAppendingString(": Sara")
```

Swift made a small improvement to Objective-C code so that it looks more Swift-style. The biggest change is made to instance creation and the style of the initialization code. The `init`, the `initWith`, and other factory methods are transformed into Swift initializers:

```
//Objective-C

- (instancetype)initWithFrame:(CGRect)frame;
+ (UIColor *)colorWithWhite:(CGFloat)white alpha:(CGFloat)alpha;

// Swift
init(frame: CGRect)
init(white: CGFloat, alpha: CGFloat)
```

The other change is made to `NS_ENUM` and `NS_OPTIONS`. They become native Swift types: `enum` and `RawOptionSetType`.

As you can see, the API looks a bit different. Because Swift strives for cleanliness, it removes word duplications from the API nomenclature. The other method calls, properties, and names, are the same as they were in Objective-C, so it should be easy to find and understand them.

What is happening behind the scenes is that Swift is generating special interface files to interact with Objective-C. You can see these Swift interface files by holding down the *command* key and clicking on the type, `NSDate` and `UIColor` in our example.

Using Swift in Objective-C

It is also possible to use Swift in Objective-C. It makes Swift very easy to adapt to an existing project. You can start by adding one Swift file, and move more functionality to Swift over time.

The setup process is much easier than that for including Objective-C in Swift. All you need to do is import Swift's autogenerated header to Objective-C. The naming convention of the files for application targets is `ProductModuleName + -Swift.h`, and for frameworks, it is `<ProductName/ProductModuleName + -Swift.h>`.

Take a look at the following examples:

```
#import "SwiftApp-Swift.h"
#import <MySwiftKit/MySwiftKit-Swift.h>
```

You can inspect the content of that autogenerated file by holding down the *command* key and clicking on it. By default, Swift classes aren't exposed for use in Objective-C. There are two ways of making Swift classes available in Objective-C:

- Mark the Swift class, protocol, or enumeration with the `@objc` attribute.

    You can mark classes, methods, protocols, and enumerations with the `@objc` attribute. The `@objc` attribute also accepts the alternative name that is used for Objective-C. When you expose a Swift class by marking it with the `@objc` attribute, it has to inherit from the Objective-C class, and the enumeration must have a raw `Int` value:

```
@objc(KOKPerson) class Person: NSObject {
   @objc(isMan) func man() -> Bool {
      ...
   }
}
@objc enum Options: Int {
   case One
   case Two
}
```

    Now, the `KOKPerson` class with the `isMan` method is available for use in Objective-C.

- Inherit from an Objective-C class, `NSObject` for example:

    When you inherit from an Objective-C class, your Swift class automatically becomes available in Objective-C. You don't need to perform any extra steps in such cases. You can also mark it with the `@objc` attribute and provide an alternative name:

```
class Person: NSObject {
}
```

Features of Swift that are not available in Objective-C

There are some features of Swift that are not available in Objective-C, so if you plan to use Swift code from Objective-C, you should avoid using them. Here is the complete list of these features:

- Structures
- Generics
- Tuples
- Enumerations
- Type aliases
- Top-level functions
- Curried functions
- Global variables
- Swift-style variadic parameters
- Nested types

Performance – meaning and key metrics

There are two key characteristics of code:

- **Code quality**: It has to be solid and flexible and have a good architecture
- **Code performance**: It has to be fast

Making the code architecture very solid and stable is the most important task, but we shouldn't forget about making it fast as well. Achieving high performance can be a tricky and dangerous task. Here are a few things that you should keep in mind while working on performance improvement:

- Don't optimize your code upfront

    There are many articles about this topic, why it's dangerous, and why you shouldn't do it. Just don't do it, and as Donald Knut says:

    "Premature optimization is the root of all evil"

- Measure first

  Firstly, don't optimize upfront, and secondly, measure first. Measure the code's performance characteristics and optimize only those parts that are slow. Almost 95 percent of code doesn't require performance optimization.

  I totally agree with these points, but there is another type of performance optimization that we should think of upfront.

Everyday code performance

The small decisions that we make every day include the following:

- What type should it be, Int or String?
- Should I create a new class for a new functionality or add to an existing one?
- Use an array? Or maybe a set?

It seems as if these don't have any impact on the application's performance, and in most cases, they don't. However, making the right decision not only improves an application's speed, but also makes it more stable. This gives higher performance in application development. The small changes that we make every day make a big impact at the end of the year.

The importance of performance

High performance is very crucial. The performance of an app is directly related to user experience. Users want to get results immediately; they don't want to wait for the view to load, see a long **Loading** indicator, or see a lagging animation.

Every year, our computers and devices become more and more powerful, with more CPU speed, memory, storage, and storage speed. Performance problems could seem irrelevant because of this, but the software complexity increases as well. We have more complex data to store and process. We need to show animations and do a lot of other things.

The first way of solving a performance problem is by adding more power. We can add more servers to handle data, but we can't update our clients' PC and mobile devices. Also, adding more power doesn't solve the code performance issue itself, but just delays it for some time.

The second, and correct, solution is to remove the issue that causes the performance problem. For that, we need to identify the problem, the slow piece of the code, and improve it.

The key metrics

There are many things that impact an application's performance and user experience. We will cover the following key metrics:

- Operations' performance speed
- Memory usage
- Disk space usage

The most important of these and the one that has the biggest impact is the **operations' performance speed**. It tells us how fast a particular task can be performed, for example, creating a new user, reading from a file, downloading an image, searching for a person with a particular name, and so on.

Summary

Swift is a powerful and fast programming language. In this chapter, you learned about many powerful features of Swift and how easy it is to start coding in Swift and integrate it into existing projects. We also covered why performance is important and what you should be thinking about when working with it.

In the next chapter, we will do more coding in Swift, and you will learn how to use all the features of Swift to make a good application architecture.

2
Making a Good Application Architecture in Swift

Swift is a high-performance programming language, as you learned in the previous chapter. You also learned that writing good code is even more important than making it high-performance code. In this chapter, we will put the all-powerful features of Swift together and create an application. We will do this by covering the following topics:

- Writing clean code
- Immutability
- Value types and immutability
- Representing the state with classes
- Representing the absence of values with optionals
- Functional programming
- Generics

Making a Swift application

The first step in creating a good application architecture is to create the application itself. We will be creating an iOS journal application used to make daily notes. We are not going to cover any iOS-specific topics, so you can use the same code and create OS X applications as well.

Go ahead! Open Xcode and create a new iOS single-view project application. Now, we are ready for coding.

First, let's create a `Person` type, for the owner of the journal, and a journal entry type. We will use the `Class` type to create both `Person` and `JournalEntry`. Both classes are very simple—just a bunch of properties and an initializer:

```swift
class Person {
  var firstName: String
  var lastName: String

  init (firstName: String, lastName: String) {
    self.firstName = firstName
    self.lastName = lastName
  }
}

class JournalEntry {
  var title: String
  var text: String
  var date: NSDate

  init (title: String, text: String) {
    self.title = title
    self.text = text
    date = NSDate()
  }
}
```

This is the minimal setup that we need for the app. Before we move forward, let's make the code better.

The differences between variables and constants

Probably, the most often used feature in all programming languages is creating and storing a value. We create local variables in functions and declare them in classes and other data structures; that's why it's very important to do it properly.

In Swift, there are two ways of creating and storing a value, as follows:

- Making it a variable:

```swift
var name = "Sara"
```

- Making it a constant:

```swift
let name = "Sara"
```

The difference between variables and constants is that a constant value can be assigned only once and can't be changed after that. A variable value, on the other hand, can be changed anytime. Here's an example:

```
var name = "Sam"
name = "Jon"

let lastName = "Peterson"
lastName = "Jakson" //Error, can't change constant after assigning
```

> The golden rule is to always declare your type as a constant (the `let` keyword in the previous example) first. Change it to a variable (the `var` keyword) only if you need it afterwards.
>
> There are some exceptions when you can't declare it as a constant, for example, when making `@IBOutles` or `weak`. Also, optional values must be declared as variables.

Using constants has many benefits over using variables. A constant is an immutable type, and we will cover all the benefits of immutability later. The two most important benefits are as follows:

- Safety (protection from unexpected value changes)
- Better performance

You should use constants both when declaring properties and as local constants in functions. We should apply this rule and change our `Person` and `JournalEntry` classes as follows:

```
class Person {
  let name: String
  let lastName: String
  ...
}

class JournalEntry {
  let title: String
  let text: String
  let date: NSDate
  ...
}
```

Usually, you will find yourself using constants more often than variables. Let's look at an example where you could think about using a variable but, in fact, a constant would be a better solution. Let's say you have created a new person in the application and now you want to display a full name with a gender prefix:

```
let person = Person(firstName: "Jon", lastName: "Bosh")
let man = true

var fullName: String
if man {
  fullName = "Mr "
} else {
  fullName = "Mrs "
}

fullName += person.firstName
fullName += " "
fullName += person.lastName
```

If you think a bit more about the problem, you will realize that `fullName` of the person should be immutable; it's not going to change, and it should be declared as a constant:

```
let person = Person(firstName: "Jon", lastName: "Bosh")
let man = true

let gender: String = man ? "Mr": "Mrs"
let fullName = "\(gender) \(person.firstName) \(person.lastName)"
```

Immutability

In the previous section, you learned how important it is to use immutable constants. There are more immutable types in Swift, and you should take advantage of them and use them. The advantages of immutability are as follows:

- It removes a bunch of issues related to unintentional value changes
- It is a safe multithreading access
- It makes reasoning about code easier
- There is an improvement in performance

By making types immutable, you add an extra level of security. You deny access to mutating an instance. In our journal app, it's not possible to change a person's name after an instance has been created. If, by accident, someone decides to assign a new value to the person's `firstName`, the compiler will show an error:

```
var person = Person(firstName: "Jon", lastName: "Bosh")
p.firstName = "Sam" // Error
```

However, there are situations when we need to update a variable. An example could be an array; suppose you need to add a new item to it. In our example, maybe the person wants to change a nickname in the app. There are two ways to do this, as follows:

- Mutating an existing instance
- Creating a new instance with updated information

Mutating an instance in place could lead to a dangerous, unpredictable effect, especially when you are mutating a reference instance type.

 Classes are reference types. "Reference type" means that many variables and constants can refer to the same instance data. Changes done to the instance data reflect in all variables.

Creating a new instance is a much safer operation. It doesn't have any impact on the existing instances in the system. After we have created a new instance, it may be necessary to notify other parts of the system about this change. This is a safer way of updating instance data. Let's look at how we can implement a nickname change in our `Person` class. First, let's add a nickname to the `Person` class:

```
class Person {
  let nickName: String
  ...

  func changeNickName(nickName: String) -> Person  {
     return Person(firstName: firstName, lastName: lastName,
               nickName: nickName)
  }
}

let sam = Person(firstName: "Sam", lastName: "Bosh",
  nickName:"sam")
let rockky = sam.changeNickName("Rockky")
```

Because we made a `sam` instance a constant, we can't assign a new value to it after changing `nickName`. In this example, it would be better to make it a variable because we actually need to update it:

```
var sam = Person(firstName: "Sam", lastName: "Bosh",
  nickName:"BigSam")
sam = sam.changeNickName("Rockky")
```

Multithreading

We get more and more core processors nowadays, and working with multithreading is a part of our life. We have GCD and NSOperation for performing work on multiple threads.

The main issue with multithreading is synchronizing read-and-write access to data without corrupting that data. As an example, let's create an array of journal entries and try to modify it in the background and main thread. This will lead to an application crash:

```
class DangerousWorker {
  var entries: [JournalEntry]

  init() {
    //Add test entries
    let entry = JournalEntry(title: "Walking", text: "I was
      walking in the loop")
    entries = Array(count: 100, repeatedValue: entry)
  }

  func dangerousMultithreading() {

    dispatch_async(dispatch_get_global_queue(
                DISPATCH_QUEUE_PRIORITY_BACKGROUND, 0)) {
      sleep(1) //emulate work
      self.entries.removeAll()
    }

    print("Start Main")
    for _ in 0..<entries.endIndex {
      entries.removeLast() //Crash
      sleep(1) //emulate work
    }
  }
}
```

```
let worker = DangerousWorker()
worker.dangerousMultithreading()
```

These kinds of issues are really hard to find and debug. If you remove the `sleep(1)` delay, the crash might not occur on some devices, depending on which thread is run first.

When you make your data immutable, it becomes read-only and all threads can read it simultaneously without any problems:

```
let entries: [JournalEntry]

let entry = JournalEntry(title: "Walking", text: "I was walking")
entries = Array(count: 100, repeatedValue: entry)
// entries is immutable now, read-only

dispatch_async(dispatch_get_global_queue(
            DISPATCH_QUEUE_PRIORITY_BACKGROUND, 0)) {

  for entry in self.entries {
    print("\(entry) in BG")
  }
}

for entry in self.entries {
  print("\(entry) in BG")
}
```

But we often need to make changes to the data. Instead of making changes directly to the source data, a better solution is to create new, updated data and pass the result to the caller thread. In this way, multiple threads can safely continue performing a read operation. We will take a look at multithreading data synchronization in *Chapter 6, Architecting Applications for High Performance*.

Value types and immutability

There are two different data types in Swift:

- Reference types
- Value types

Let's take a look at these.

Reference types

A class is a reference type. When you create an instance of a reference type and assign it to a variable or constant, you are not only assigning a value but also a reference that points to the value, which is located somewhere else (actually it is located in the heap memory). When you pass that reference to other functions and assign it to other variables, you are creating multiple references that point to the same data. If one of those variables changes the data, that change will reflect in all other variables as well. Here's an example that shows this:

```
let person = Person(firstName: "Sam", lastName: "Jakson")
let a = person, b = person, c = person
```

The following diagram shows what the memory for this code would look like:

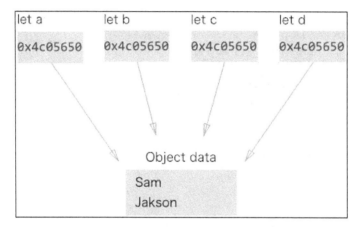

All four constants would refer to the same object. The danger in this architecture is that if one of those constants updates a piece of instance data, every other constant would get updated as well. Here's an example that shows this:

```
a.firstName = "Jaky"
b.firstName // Jaky
```

Sometimes, this can be desirable behavior, for example, when many variables are referencing the same window object. There should be only one window object, and changes made in one place should be reflected in others as well.

Value types

A structure, on the other hand, is a value type. When you create an instance of a value type and assign it to a variable, you are assigning actual data. When you pass that instance to other functions and variables, you are passing a copy of that value. The copy is made automatically. You may think that copying values would have a negative impact on performance, but in reality, value types give higher performance than reference types. Value types are smart enough to optimize data copying only when it's required (when data is being modified).

If we make our `Person` type a `Structure` type, the same code example would look like this:

```
let person = Person(firstName: "Sam", lastName: "Jakson")
let a = person, b = person, c = person
```

The memory for this structure's constants would look like this:

let a	let b	let c	let d
Sam	Sam	Sam	Sam
Jakson	Jakson	Jakson	Jakson

The advantage of this architecture is that your code components are isolated and not dependent on each other.

One big difference between reference types and value types is explained as follows: when you create a constant as a reference type, you are making a constant reference (which means that you can't change it to point to another instance). But you can change the data in the instance itself, as we did in the example by updating the `firstName` of the person.

When you make a constant of a value type, you get a constant value that can't be changed.

The power of structures

If you look more closely at the Swift standard library type definition, you will discover that most of the types are implemented as a structure, such as `struct Int`, `struct String`, `struct Array`, and others.

The structure is not only a simple and fast data structure, but also a very powerful one. Structures can have methods, properties, and initializers, and they can conform to protocols. When you're designing your entities in the application, try to use structures for your data model, and structures are preferred in general. Now we will apply this advice and change our types to use a structure instead of a class:

```
struct Person {
   let firstName: String
   let lastName: String
   let nickName: String

   func changeNickName(nickName: String) -> Person  {
     return Person(firstName: firstName, lastName: lastName,
                nickName: nickName)
   }
}

extension Person {

   init(firstName: String, lastName: String) {
     self.init(firstName: firstName, lastName: lastName,
              nickName:"")
   }
}
```

The first difference is that we changed the `class` keyword to `struct`. The second one is more interesting—we removed the `init` method. If you don't define an initializer, a structure provides a default member-wise initializer. A member-wise initializer takes all the properties of the structure. If you need an extra initializer in addition to a default member-wise one, you can create it in an extension. In this way, you would have two initializers:

```
Person(firstName: "Sam", lastName:"Niklson", nickName: "Bigsam")
Person(firstName: "Petter", lastName: "Hanson")
```

Representing the state with classes

When designing a data model in the application, use value types. The value types should be:

- Inert
- Isolated
- Interchangeable

The value type shouldn't behave and it shouldn't have side effects. The operation on the data should go to the value layer. You can learn more about designing data models with value types in the presentation *Controlling Complexity in Swift* by Andy Matuschak at `https://realm.io/news/andy-matuschak-controlling-complexity/`.

Classes, on the other hand, can have a behavior and a state. An action of creating a new `JournalEntry` is a behavior, for example, and it should implemented in the class type. The current user's `JournalEntry` list is a state, and this should also be stored in the class type:

1. First, what we do is create a `Journal` data model as a value type. It contains data and operations to work with that data (it has the `addEntry` method, which creates and adds new entries to the journal):

```
struct Journal {
    let owner: Person
    var entries: [JournalEntry]

    mutating func addEntry(title: String, text: String) {
        let entry = JournalEntry(title: title, text: text)
        entries.append(entry)
    }
}

extension Journal {
    init(owner: Person) {
        self.owner = owner
        self.entries = []
    }
}
```

2. The next step is to create a controller entity as a reference type that will hold the current journal state in the application and handle the action of adding a new entry:

```
class JournalController  {
  var journal: Journal

    init(owner: Person) {
        self.journal = Journal(owner: owner)
    }

    func addEntry(title: String, text: String) {
        journal.addEntry(title, text: text);
    }
}
```

Representing the absence of values with optionals

Let's go back to the past and see how the absence of a value is represented in Objective-C, as an example. There isn't a standard solution for representing the absence of a value for both reference and simple value types. There are two different ways:

- `nil`
- `0`, `-1`, `INT_MAX`, `NSNotFound`, and so on

For reference types, Objective-C uses the `nil` value to represent that a variable doesn't have a value. It points to nowhere.

For value types, there is no such value as `nil` and it is not possible to assign `nil` to an integer variable. To do that, Objective-C (and not only Objective-C but also C, Java, and many other languages) uses a few special values that are unlikely to be the result of a particular operation. For example, the `indexOfObject` method of `NSArray` would return `NSNotFound`.

 NSNotFound is just a constant and its value is equal to `NSIntegerMax`, whose value, in turn, is `2147483647`.

Swift uses an optional to represent the absence of a value in a common way for both value and reference types. An optional is a way of annotating that the value could be missing. You can declare a type as optional in two ways:

- Using the optional keyword, that is, `Optional<Type>`
- By adding a question mark to the end of the type, that is, `Type?`

 `Type?` is the preferred way to declare an optional type.

To represent a missing value, you can simply assign `nil` to an optional, as shown in the following example:

```
var view: Optional<UIView>
var index: Int?

view = nil
view = UIView()

index = 10
index = nil
```

Optional and non-optional types

In Objective-C both optional and non-optional values are represented by the same type, for example, NSInteger, NSString *. By looking at the source code and method definition, there is no way to say whether a method can return nil or NSNotFound or not:

```
-  (NSUInteger) indexOfObject: (id) anObject;
+  (instancetype) stringWithString: (NSString *) string;
```

 In Xcode 6.3, we have new Objective-C annotations, nullable and nonnull, that allow us to specify whether nil can be passed or not. These annotations were added after Swift's release to provide better Objective-C integration with Swift.

Swift is stricter about this. It has both optional and non-optional types. Two examples of non-optional types are Int and String. This means that you can't assign nil to an Int variable or pass nil to a function with the Int parameter type. Optional types, on the other hand, allow you to use nil:

```
var index: Int?
var number: Int = 10

index = nil // Ok
number = nil // Error

func indexOfObject(object: Any) -> Int?
func stringWithString(string: String?) -> String?
```

This strict rule makes the code's intention really clear. From the API, you see that to call an indexOfObject function, you need to pass a non-optional argument, and it could return nil as a result.

Safe nil handling

The other problem is in trying to access nil values. If you have been programming in C, Java, or Objective-C, you must have faced one of the NullPointerException exception or the NSInvalidArgumentException exception.

In general, it is unsafe to access uninitialized memory. For example, passing nil to the initWithString method in Objective-C would result in an NSInvalidArgumentException exception, and maybe an application crash:

```
[NSString initWithString: nil] - crash
```

The sad part is that Objective-C didn't check the difference between `String * ` type and `nil`.

Optionals in Swift not only make it clear about the ability to use `nil`, but also make it very safe to work with and avoid crashes.

Using optionals

Now that you've understood the background on why optionals were invented, let's go ahead and use them in our application. We have our `JournalEntry` entity, and let's say the user can add a location where this entry was created (this is an optional feature; some entries will have it and some will not). We need to create a new type to store the geographic location, and add a new `Optional` property to our `JournalEntry` entity:

```
struct Location {
   let latitude: Double
   let longitude: Double
}

struct JournalEntry {
   var location: Location?
...
}
var entry = JournalEntry(title: "Walking",
                         text: "I was walking in the loop")
let location = Location(latitude: 37.331686,
                        longitude: -122.030656)
entry.location = location
```

Optional variables are assigned a `nil` value by default, and because of this, we don't need to make any more changes to our `init` methods (all the properties have been provided with a value).

> The `var: Int? = nil` is the same as `var: Int?`. Don't assign a nil value if you are declaring an optional variable.

This is much more interesting when you try to access an optional value. An optional is like a closed box with something inside. To get the value out of it, you have to open it first. To check whether an optional has a value inside it, use the if and == nil or != nil comparison operator. To get actual data out of the box, you need to unwrap it using ! sign:

```
if entry.location != nil {
    showLocation(entry.location!)
} else {
    //locationNotAvailable
}
```

This, however, is not the optimal way of getting a value from an optional. A better way is to use the optional binding operator, which checks whether the optional has a value, and unwraps its value at once. The syntax is if let unwrappedValue = optional:

```
let location = entry.location
if let location = location {
    showLocation(location)
} else {
    //locationNotAvailable
}
```

First, we extract the location to a local constant, and it has the Optional<Location> type. Next, we apply the optional binding operator and get a value from the options to a location constant. The constant name for an optional value is the same as the optional name (location in this example). This technique is called **name shadowing**. When you use a location as an argument for the showLocation function, you are using the unwrapped constant value.

> When using the optional binding operator, use name shadowing. This makes the code much more readable. Here is how it would look without name shadowing:
>
> ```
> if let location = maybeLocation {
>     showLocation(location)
> }
> ```
>
> Adding different names for the optional and its unwrapped value (maybeLocation and location in our example) makes the code more confusing.

There is also another type of optionals available in Swift—implicitly unwrapped optionals. You declare them with `Type!`, for example, `Int!`. An implicitly unwrapped option is one that does not require checking of whether values exist inside an option, but allows you to access the data as if it were not the optional type. Here is an example:

```
var name: String! = "Jon"
print("My name is" + name)

name = nil
print("My name is " + name) // Crash
```

Using implicitly unwrapped optionals is unsafe and not recommended. There are very few situations where you should use them. They are mostly used to interact with the Objective-C API. This is because many types from Objective-C are transformed into Swift as implicitly unwrapped optionals.

 Avoid using implicitly unwrapped optionals.

Wrapping up on optionals

If the idea of optionals is new to you, don't be afraid; you will get used to working with them, and you will like them. Here are few small notes that you should remember from this topic:

- Don't use implicitly unwrapped optionals wherever possible—almost never!
- Check whether an optional has a value before accessing it
- Use optional binding and shadowing of the optional variable name to access a value
- Design your API's intentions to be clear with optional and non-optional types

Functional programming

In the functional programming paradigm, a function has a type and it is treated in the same way as other types, such as `Int`, `String`, and `Class`. The function can be assigned to a variable, passed as an argument to another function, and returned from a function as a result type. The main goal is to split the code into small, standalone functions. The perfect function has no side effects and operates only with arguments that were passed to it.

In the functional programming style, you describe *what you want to do* and not *how you want to do it*.

Functional programming is great for data transformation and data manipulation. This is because you are able to split code into smaller parts. You can often reuse some routine boilerplate code.

Function types

Every function has a type. The type of a function consists of its parameter type and return type. Now, we will create a few functions with different types and perform some operations:

```
func hello() {
    print("Hello")
}

func add(x: Int, y: Int) -> Int {
    return x + y
}

func subtract(x: Int, y: Int) -> Int {
    return x - y
}

var hi: () -> () = hello
var mathOperation: (Int, Int) -> Int

mathOperation = add
mathOperation(10, 11) // 21
mathOperation = subtract
mathOperation(10, 11) // -1

hi()
mathOperation = hello // Error, wrong types
```

The `hello` function has the `() -> ()` type. It takes nothing and returns nothing. The `add` and `subtract` functions have a different type: `(Int, Int) -> Int`. In the preceding example code, we assigned functions to the `hi` and `mathOperation` local variables.

It is not possible to assign the `hello` function to the `mathOperation` variable because they have different types.

Splitting the code

Because we can pass one function to another, we can split the code into the actual logic and the routine work. Let's implement a very common operation. The task is to double every element in the array. In imperative programming, this task would translate into iterating over every element in the array, doubling every element, and saving the result in a new array. In the end, the result array is returned with all its elements doubled:

```
let numbers = [1, 2, 3]

func doubleNumbers(array: [Int]) -> [Int] {

  var result = [Int]()
  for element in numbers {
    result.append(element * 2)
  }
  return result
}

let result = doubleNumbers(numbers) // [2, 4, 6]
```

The problem with this code is that there is only one line that actually does the work, that is, element * 2. It can't be reused because this logic is hardcoded into the function body. What if we want to triple numbers or do some other transformation? Here is how this task is implemented in the functional way:

```
func transform(array: [Int], f: Int -> Int) -> [Int] {

  var result = [Int]()
  for element in array {
    result.append(f(element))
  }
  return result
}
```

The only difference here is that the transform function takes a transformation function as an argument. The transformation function does all the routine work, iterating over an array, but it leaves the actual transformation logic to be performed by the function that you passed as an argument. In this way, you can pass different functions to the transform function:

```
func double(x: Int) -> Int {
    return x * 2
}
```

```
func triple(x: Int) -> Int {
    return x * 3
}

let result = transform(numbers, f: double)
let result = transform(numbers, f: triple)
```

The closure expression

A closure expression is an inline, unnamed, and self-contained block of code. You can think of a closure expression as a function without a name; it also takes parameters and has a body and a return type. You can use a closure instead of a function if they have the same type.

The general syntax for a closure expression is as follows:

```
{ (parameter name: type) -> return type in  body }
```

Let's refactor our `transform` function to use a closure instead. Here is the result:

```
transform(numbers, f: { (x: Int) -> Int in
  return x * 2
})
```

Because closure expressions are designed to be used inline, they have many syntax optimizations for making them clean and clear. Here are some examples of these optimizations:

- Type inference
- Implicit return type
- Shorthand argument name
- Trailing closure syntax

Type inference

Thanks to type inference, you don't need to specify the parameter type and return type, as shown here:

```
transform(numbers, f: { x in
  return x * 2
})
```

 As a general rule, you should avoid specifying types whenever it is possible.

Implicit return type

A closure with a single-expression body implicitly returns the result of that expression. The `return` keyword can be omitted in such cases:

```
transform(numbers, f: { x in x * 2 })
```

In this example, we can't omit the `return` keyword because there is more than one expression:

```
transform(numbers, f: { x in
  let result = x * 2
  return result + 10
})
```

Shorthand argument names

You can omit argument names from a closure expression. In such cases, Swift provides a default name for every argument. This name consists of the $ sign and the argument index, for example, $0, $1, $2, and so on:

```
transform(numbers, f: { $0 * 2 })
```

 Shorthand argument names are preferred for very short closure expressions in which an argument is used once or twice. In other cases, give your argument a descriptive name.

Trailing closure syntax

When a function's last argument is a closure, you can write the closure expression outside the function call:

```
transform(numbers) { $0 * 2 }
```

You can use all closure syntax with trailing closures:

```
transform(numbers) { x in x * 2 }
```

If a function has only one argument and it is a closure, you don't need to specify empty parentheses for the function call:

```
func map(function: Int -> Int) -> [Int] {
  ...
}
map() { $0 * 2 }
map { $0 * 2 }
```

The standard library

Swift's standard library has many functions and methods that accept other functions. Here are a few methods of `SequenceType` that you should know and use:

- `map`
- `reduce`
- `filter`
- `sort`

The map method

The `map` method applies a `transform` function to every item and returns the new resultant collection. This process is called mapping, where the values A to B are mapped:

```
func map(transform: (Int) -> String) -> [String]
```

 Swift's standard library uses the `generic` function, but in the following example, they have been changed to actual types to provide simpler examples. Here is the actual definition of the `map` function:

```
func map<T>(@noescape transform: (Self.Generator.
Element) -> T) -> [T]
```

The `map` method does exactly the same job as our `transform` function. So, you should be using `map` instead:

```
let result = numbers.map(double)
let result = numbers.map { $0 * 2 }
```

map for optionals

The `Optional` type also has the `map` method, but here it works differently. It takes a function that maps an optional value, if it exists, to another value:

```
func map<U>(@noescape f: (Wrapped) -> U) -> U?
```

The body of this `map` method would look like this:

```
func map(f: (Wrapped) -> Double) -> Double? {
  switch self {
    case .None: return nil
    case .Some(let x): return f(x)
  }
```

```
    }

    let number: Int? = 10
    let res = number.map { Double($0) * 2.3 }
```

Using `map` with optionals can make your code cleaner. Consider the following example, which uses the `map` function and manual unwrapping optionals:

```
    // Using the map function
    let doubled = number.map(double)

    // Optional binding
    let doubled: Int?
    if let number = number {
      doubled = double(number)
    } else {
      doubled = nil
    }
```

The reduce method

The `reduce` method takes the initial value and the `combine` function. It aggregates the result by calling the `combine` function for every element in the sequence. The `combine` function takes the returned value of a previous call of the `combine` function itself or an initial value of the first call and an element of a collection:

```
    func reduce(initial: Double, combine: (Double, Int) -> Double) ->
      Double
```

The simplest use case of a `reduce` function would be to calculate the sum of a few elements. Its implementation looks like this:

```
    {
      var result = initial
      for item in self {
        result = combine(result, item)
      }
      return result
    }

    let sum = numbers.reduce(0) { acc, number in acc + number }
```

You can make this code cleaner by using closures, shorthand argument names, or operator functions:

```
numbers.reduce(0) { $0 + $1 }
numbers.reduce(0, combine: +)
```

 The + operator is defined as an operator function and can be used in every place where a function is expected.

```
infix operator + {
    associativity left
    precedence 140
}
func +(lhs: Int, rhs: Int) -> Int
```

The filter method

The `filter` method filters out elements from the source collection by asking the `includeElement` function what elements to keep. The `includeElement` function is called for every element in the source collection, and returns a boolean value that indicates whether the element should be kept or removed:

```
func filter(includeElement: (Int) -> Bool) -> [Int]
```

The implementation looks like this:

```
{
  var filtered = [Int]()
  for item in self {
    if includeElement(item) {
      filtered.append(item)
    }
  }
  return filtered
}

let evenNumbers = numbers.filter { $0 % 2 == 0 }
```

Functional programming is a very big topic. If you are interested, you can read more about it in *Functional Programming in Swift* by Chris Eidhof, Florian Kugler, and Wouter Swierstra. You can get it from http://www.objc.io/books/.

Generics

Generics are a way of writing generic, reusable code without specifying a type. You can write a `generic` function that may not be limited to one type. It's possible to create `generic` functions as well as `generic` types that add type restrictions. You have used `generic` types in this book even without noticing it.

The main idea behind generics is that instead of specifying a type, you use a generic type placeholder. Generics are a great tool for removing code duplication and making code reusable.

The first step is to identify the code that can be generic. The best way to do this is by asking, "Is this functionality limited only to this type or not?" If you realize that it is not, you should consider making it generic.

Make functions generic only if you need to do so and if you are going to use them with different types. Making them generic could have a slightly negative impact on performance.

Let's create our first simple generic function. Our `printMe` function can work only with integers as of now, but it will be great to make it work with all types:

```
func printMe(x: Int) {
    print("Me - \(x)")
}
```

To get a generic function or type, you need to specify a generic type parameter in the angle brackets (`<T>`) and use that type instead of the actual type:

```
func printMe<T>(x: T) {
    print("Me - \(x)")
}

printMe(10.0)
```

The naming convention for type parameters is camel case. In simple cases, where the `generic` type doesn't have any special meaning, use the single-character name `T`. In complex situations, you can give descriptive names, such as `Key` or `Value`.

Generic functions

The `transform` function that we wrote is a great candidate for a `generic` function. It doesn't perform any computation that requires a specific type. The only thing we need to do is use a placeholder type name for the array type and the transform function instead of the `Int` type:

```
func transform<T>(array: [T], function: T -> T) -> [T] {

  var result = [T]()
  for element in array {
    result.append(function(element))
  }
  return result
}
```

Now we can use our transform function with any type:

```
let numbers = [1, 2, 3]
let increasedNumbers = transform(numbers) { $0 + 1}

let names = ["Jon", "Sara", "Sam"]
let formattedNames = transform(names) { "Name: " + $0 }
```

Type constraints

You can't perform any operations with the variable of the generic type `T` because Swift doesn't know anything about that type. If you try to compare two arguments of type `T`, Swift will show the following error: **Could not find and overload '<' that accepts the supplied arguments**:

```
func minElem<T>(x: T, _ y: T) -> T {
  return x < y ? x : y
}
```

The comparison operator, <, is defined in the comparable protocol. We need to specify that our generic type `T` should conform to the comparable protocol. With a type constraint, you can specify that a type must conform to a protocol or inherit from a class. To do that, you list constraints after the colon (:) in the generic name definition:

```
<type parameter : constraint >

func minElem<T : Comparable>(x: T, _ y: T) -> T {
  return x < y ? x : y
}
```

Now our `minElem` function can work with any type that conforms to the comparable protocol, such as `Int` and `String`:

```
minElem(10, 20)
minElem("A", "B")
```

The great thing about making `minElem` a generic function with the `constraint` protocol is that it is not limited to only working with existing types. We don't need to make any changes to make it work with a new type. Let's say we want to find the smallest `JournalEntry` entity. All that we need to do is make sure that it conforms to the comparable protocol.

The comparable protocol requires two function operators to be implemented in your type: == and <:

```
func ==(lhs: Self, rhs: Self) -> Bool
func <(lhs: Self, rhs: Self) -> Bool
```

Let's say we want to find the smallest `JournalEntry` entity. All that we need to do is make sure that it conforms to the comparable protocol:

```
extension JournalEntry : Comparable {
}

func == (lhs: JournalEntry, rhs: JournalEntry) -> Bool {

    return lhs.title == rhs.title &&
      lhs.text == rhs.text &&
      lhs.date == rhs.date
}

func < (lhs: JournalEntry, rhs: JournalEntry) -> Bool {
    return lhs.text < rhs.text
}
```

Conform to a protocol in a type extension. In this way, you can split the code into functional sections.

When you conform to protocols in the type declaration, the type declaration becomes hard to read and contains too much information:

```
struct JournalEntry : Comparable, Hashable,
    CustomStringConvertible {

    . . .

}
```

Now, we can create two `JournalEntry` entities and call a `minElem` function. The `minElem` function will use the `<` operator function to compare two journal entries:

```
let walking = JournalEntry(title: "Walking",
        text: "It was a great weather")
let goal = JournalEntry(title: "Read", text: "Read a book")
let smaller = minElem(walking, goal)
```

The generic type and collections

Another great use case for generics is to make a generic type. `Array`, `Dictionary`, and `Set` are implemented as generic types:

```
struct Array<T> ...
struct Dictionary<Key : Hashable, Value> ...
struct Set<T : Hashable> ...
```

This gives us the ability to store any type in collections and make them single-type collections. This means that we can't store the wrong type in them:

```
var numbers = [1, 2, 3] // [Int]
numbers.append(10)
numbers.append("Name") //Error, Can't add String to [Int] array
```

You can make your own your custom generic types. The rules are the same as for declaring a generic function; you specify a generic type in angular brackets, and use it everywhere as a type name. As an example, we can make our own simple generic stack like this:

```
struct Stack<T> {
  private var items: [T]

  mutating func push(item: T) {
    items.append(item)
  }

  mutating func pop() -> T {
    return items.removeLast()
  }

  init() {
    items = []
  }
}

var s = Stack<Int>()
```

```
s.push(10) // 10
//s.push("Name") // Error
s.push(4)   // 10, 4
s.pop()     // 10
```

Safety

Swift is designed for safety. It eliminates many issues of compile time. Here is a list of things that Swift handles for you:

- **Type safety**: Swift is very strongly typed language. If a function has the `Int` parameter, you must pass `Int` as an argument when you call it. This rule also applies to operators. Swift doesn't allow use of the wrong type:

```
func increase(x: Int) -> Int {
   return x + 1
}

let x = 10
let percent = 0.3
let name = "Sara"

x + name //Error, can't apply + operator for Int and String
x * percent //Error, can't apply * to Int and Double
Double(x) * percent // 3

increase(x) // 11
increase(percent) // Wrong type
increase(name) // Wrong type
```

- **Variables must always be initialized before use**: Accessing non-initialized memory is a dangerous operation. Swift handles this problem in a very nice and safe way. It doesn't compile when you try to do that:

```
var y: Int
//y + 10 //Error, variable 'y' used before being initialized
y = 1
y + 10
```

  Constant values can't be changed after they have been set, but you can declare a constant without setting an initial value and set it later:

```
let z: Int

if y == 2 {
   z = 10
```

```
} else {
  z = 0
}
z + 10
```

If you remove the `else` case, the Swift compiler will show an error, because in these cases, z won't be initialized when y `!= 2`.

- **Safe nil handling**: As you have seen already Swift has an optional type for safe nil handling and absence-of-value handling.

Dangerous operations

There are still situations where we should be careful, because we could make an error and cause the application to crash. Here's a list of these situations:

- **Implicitly unwrapping optionals**: Unwrapping an optional (the `!` operator) is a potentially dangerous operation. You should do it only when you have verified that the optional has a value. It is, in fact, better to use optional binding.

  Using the implicitly unwrapped optional is also a very dangerous operation. They behave like non-optional types (which do not require unwrapping before accessing a value), but they cause a crash when used with the `nil` value:

  ```
  var x: Int?
  x! + 10 // Crash! Unwrapping optional that does not have value.

  var y: Int!
  y + 10 // Crash! Implicitly unwrapped optional has nil value.
  ```

- **Type casting**: There are situations where you would like to store any object of a base class object and check later on whether that object actually has a specific type. You can safely check for the type of an object with the `is` keyword:

  ```
  var view: UIView = UIImageView()
  if view is UIImageView {
    print("yes")
  } else {
    print("no")
  }
  ```

Often, you not only need to check whether the variable is of a certain type, but also need to cast it to the corresponding type. You can do this in two ways: safe and unsafe. You should always use the safe way.

Unsafe casting is very similar to unwrapping optionals. It tries to do casting without checking whether it's possible, and this could lead to a crash:

```
let imageView = view as! UIImageView
```

Safe casting is like optional binding. First, it checks whether a view is actually a UIImageView type, and then it performs casting. Finally, it saves the cast result in a view constant:

```
if let view = view as? UIImageView {
  view.image = UIImage(named: "image")
}
```

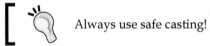 Always use safe casting!

- **Unsafe types and operations**: You will find many types and methods in Swift's standard library that start with the word Unsafe*. Those are particularly dangerous operations, and you can understand this from their names. Usually, you would use unsafe types to work with C functions. Let's see an example of the count function in C, which takes the pointer to Int as an argument:

```
int count(int *a); // C function
```

The count function in C would be available in Swift with this type:

```
count(a: UnsafeMutablePointer<Int32>)
```

You don't need to make an UnsafeMutablePointer<32> variable. You can pass an Int32 Swift variable as an in-out parameter by reference:

```
var x: Int32 = 10
count(&x)
```

You can also directly manipulate the pointer's memory, but this is a very dangerous operation and it should be avoided:

```
let pointer = UnsafeMutablePointer<Int>.alloc(1)
pointer.memory = 10
pointer.memory // store Int value - 10
pointer.dealloc(1)
```

Avoid using unsafe types. The only use case is for interacting with the C function and core foundation.

- **Accessing arrays' elements**: Although working with arrays is safe, it has an unsafe operation that you should be aware of—accessing an element beyond its range. As an example, let's create an array with three elements. Swift still allows us to try to access an element at index 10, which will lead to a crash. Swift does check the array bounds and does not allow us to use or update the memory outside the array. This prevents memory corruption issues, but it does not prevent the application from crashing:

```
let numbers = [1, 2, 3]

numbers.count
numbers[1]
numbers[10]  //Crash
```

For safety, check bound arrays before accessing an element:

```
if numbers.count > 10 {
  numbers[10]
}
```

Summary

In this chapter, we covered some of the most important and powerful features of Swift. Now you should be confident to use them. Also, this chapter gave you some advice on how to use these features and create solid applications.

In the next chapter, you will learn different debugging techniques that will help identify slow code. As you have already learned, it's very important to identify what causes performance issues before doing any optimization.

3

Testing and Identifying Slow Code with the Swift Toolkit

The process of application development can usually be split into three phases:

- Trying out new ideas
- Implementing code and checking whether it works correctly
- Measuring the performance of the result obtained

The first phase involves trying out a new idea. Let's say you would like to implement a sorting algorithm and you want to quickly prototype a solution.

In the second phase, you would actually implement the solution and check whether it is working correctly. In this chapter, we will cover how we can test and check whether a solution is implemented correctly.

The third and final phase involves measuring the performance of the software created. You would do this when you have developed enough code to test, or if you see bad performance characteristics while developing.

REPL

REPL stands for **read-eval-print-loop**. Swift REPL is an interactive Swift code interpreter that executes code immediately. To launch Swift REPL, open Terminal. app and execute this command:

```
$ xcrun swift
```

Now, you can start typing Swift code and see the result. A nice thing about evaluating code in REPL is that if you make an error that would eventually stop the application execution if you compile and run it, you can still continue evaluating the code and preserve all of the progress. Let's play around and try this code in Swift REPL:

```
let a = 10
let b = a + "c"
let b = a + 10
```

```
➜  ~  xcrun swift
Welcome to Apple Swift version 2.0 (700.0.57 700.0.72). Type :help for assistance.
  1> let a = 10
a: Int = 10
  2> let b = a + "c"
repl.swift:2:11: error: binary operator '+' cannot be applied to operands of type 'Int' and 'String'
let b = a + "c"
        ~ ^ ~~~
repl.swift:2:11: note:  overloads for '+' exist with these partially matching parameter lists: (Int, Int),
  (String, String), (Int, UnsafeMutablePointer<Memory>), (Int, UnsafePointer<Memory>)
let b = a + "c"
          ^

  2> let b = a + 10
b: Int = 20
  3>
```

Writing code in the REPL console is not as convenient as in the modern Xcode IDE, but being familiar with it is a useful skill. On top of the Swift REPL, Apple has built more powerful tools, such as Playgrounds, which has a nice source code editor and flexibility of Swift REPL.

Playgrounds

Playgrounds is a powerful tool for trying out code and getting the result. As its name suggests, it's a place to play. In a playground, Swift code is evaluated immediately, which is the same as in REPL. You can create a new playground by going to **File | New | Playground**. Enter the file name and create it.

A playground consists of two parts, which are shown in the next screenshot:

- Editor
- Result

Almost all the code examples shown in this book were created in playgrounds. As an example, let's create an array and play with it. We can apply filter and map functions and print the count of objects in an array:

You will see the result of the evaluated code appearing as you type. If you move the cursor to the one of the lines in the result section, it will become highlighted and two buttons will appear:

- **Quick Look**
- **Show / Hide result**

Quick Look will display more details about the executed operation. This functionality is particularly interesting for functions. As an example, if you click on the `filter` function, you can see the result of every iteration; the results are `true` or `false` values. If you show the details of a function that has a numerical result, a `map` function, for example, it can show a nice graph.

Show / Hide Result allows you to add a **Quick Look** result directly to the playground editor. It will always be visible and refresh the data unless you hide it.

Showing the Result View of a function with a numerical return type is very useful when you are working with the math functions and you want to see the results in a visual representation. As an example, let's show the result of the sin and pow functions, as you can see in the following screenshot:

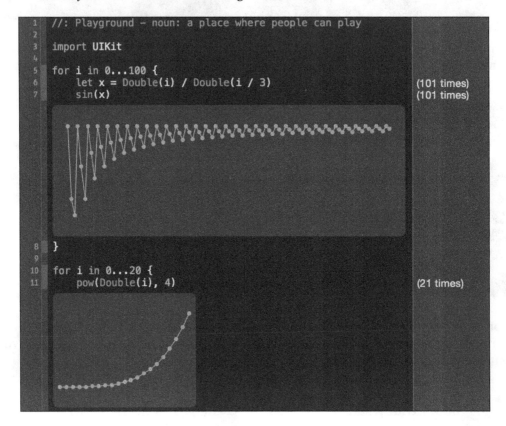

The other useful use case is for displaying the result of an algorithm. We will create a selection sort algorithm. It is often very useful to see the result of the algorithm at every step. You can easily inspect it in a playground:

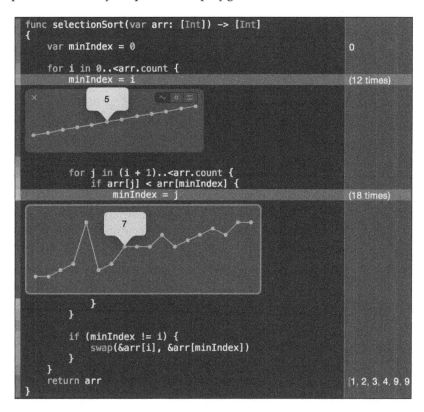

Interactive documentation

The other use case for playgrounds is in making interactive documentation. You can add markup-formatted text to the playgrounds. In this way, you can combine interactive code examples that run in the playground with rich-formatted text descriptions.

The markup syntax is based on the well-known Markdown syntax:

- **Heading**: `# Heading`
- **Strong**: `**Bold Text**` or `__Bold Text__`
- **Inline code**: `` `Int` ``

The complete Markdown documentation can be found at `http://daringfireball.net/projects/markdown/syntax`.

There are two markup text styles in Playgrounds:

- **One line**: One line markup text style is as follows:

  ```
  //: Markup text
  ```

- **Multiline**: The multiline markup text style is as follows:

  ```
  /*:

     Markup text

  /*
  ```

Let's now add some markup text to playgrounds:

```
//: # Array
//: Arrays is an ordered collection. [Read more here](https://developer.
apple.com/library/ios/documentation/Swift/Conceptual/Swift_Programming_
Language/CollectionTypes.html)
//: Arrays operations:
//: * Sort
//: * Map
//: * Etc.
```

The markup text can be presented in a playground in two modes, raw and rendered:

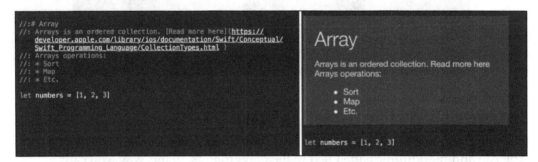

To switch between raw mode and rendered mode, go to **Editor | Show Rendered Markup / Show Raw Markup**. You can also switch between these modes by enabling the **Render Documentation** checkbox in the **Playground Settings** section in the **Utility** panel, as shown here:

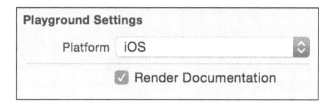

The full documentation of the Playgrounds markup format can be found at `https://developer.apple.com/library/ios/documentation/Xcode/ Reference/xcode_markup_formatting_ref`.

The folder structure

The playground file is actually not a simple file but a package that contains multiple files. You can explore its full contents by opening it in **Finder**. Right-click on **Show Package Contents**.

In Xcode, you can expand a playground file in **Project Navigator**. It contains three items:

- **Source**: A folder for additional Swift source files for the Playground
- **Resources**: A folder for additional resources for Playground such as images, text files, and other things
- **Pages**: A collection of playground files of a parent playground

The source folder

By splitting a playground's source code into several Swift source code files, you can make the playground cleaner and faster. Every time you make a change in the playground, it will rerun the entire code in that playground. The more code you put into the playground, the slower it gets. The Swift source files in the `source` folder don't rerun every time you make a change in the playground file; they rerun only when you make a change to the content of that file. As an example, when we were playing with `Person`, it's the perfect use case for adding a `Person` type to a separate Swift source file in the `source` folder.

 The Swift files in the source folder are compiled into the framework. In the framework, you have to mark your types and functions with a `public` keyword to make them visible outside the framework, in our case in the playground.

Resources

By including assets in the `Resources` folder, you can reference them in the playground. The simplest example would be adding a `circle.png` image file:

```
let circle = UIImage(named: "circle.png")
```

The files in the `Resources` folder are available via `NSBundle.mainBundle`. Let's create the same circle, `UIImage`, but this time with the `NSBundle` API:

```
if let path = NSBundle.mainBundle().pathForResource("circle", ofType:
"png") {
  let cicrcle2 = UIImage(contentsOfFile: path)
}
```

Pages

A playground file can contain many sub-playground files, called pages. A page is a fully functional playground file with its own sources and resources folders. A playground can contain many pages. To add a new page, go to **File | New | Playground Page**, or simply right-click on the playground file and select **New Playground Page**.

Playground pages are great for splitting content into separate parts, such as a page or section of a book.

For easy navigation among the pages in a playground, there is a page navigation markup. You can jump to the first, last, next, previous, or any specific page. The page navigation markup consists of two parts: [Visible text] and (page-link). Let's see some examples of this markup:

```
First and last pages links
//: [First Page] (@first)
//: [Last Page] (@last)


Next and previous pages links
//: [Next] (@next)
//: [Previous] (@previous)


Page specific links. Use the same page name as a link. The space must
be changed to "%20"
//: [Type Safe] (Type%20Safe)
//: [Optionals] (Optionals)
```

Many code examples in this book are created using playground pages.

XCPlayground

`XCPlayground` is a module specially created for working with playgrounds. It's a very small module, with four main functions:

- `XCPCaptureValue`
- `XCPShowView`
- `XCPSetExecutionShouldContinueIndefinitely`
- `XCPSharedDataDirectoryPath`

Let take a quick look at these functions. The results of all of these functions are shown in the Playground timeline. To see it, go to **View | Assistant Editor | Show Assistant Editor**, and select the timeline for the current playground.

`XCPCaptureValue` allows you to manually capture a value and display it in the timeline view. In this way, you can create your own graph results:

```
for i in 0...100 {
  let r = arc4random_uniform(100)
  XCPCaptureValue("random", value: r)
}
```

`XCPShowView` displays a view in the playground timeline:

```
let frame = CGRect(x: 10, y: 10, width: 100, height: 100)
let view = UIView(frame: frame)
view.backgroundColor = .redColor()

XCPShowView("View", view: view)
```

`XCPSetExecutionShouldContinueIndefinitely` is very useful for asynchronous code execution in a playground. It tells playground to keep executing its run loop infinitely after the last instruction completes, and in this way, we can wait for asynchronous callbacks.

`XCPSharedDataDirectoryPath` returns a path to the directory that is shared between all the playgrounds. In this way, you can save and share data between playgrounds and between each playground run.

LLDB

LLDB is a high-performance command-line debugger. It is available in Xcode. The easiest way to start it is to set a breakpoint and run the application. In the Xcode debug area view, you will find a console in which you can execute LLDB commands. Because we made an iOS app, we will set a breakpoint in the AppDelegate's `didFinishLaunchingWithOptions` method.

To print the content of a variable, we can use the `p` LLDB command. Just run `p` with the variable name, for example, `p name`, as shown here:

```
24    func LLDBTutorial() {       (lldb) p name
25        var name = "Sam"         (String) $R0 = "Sam"
26  ▶ |                            (lldb) |
27    }
```

LLDB is a very powerful debugger. You can read more about the LLDB debugger at `www.objc.io/issues/19-debugging/lldb-debugging/` and `https://developer.apple.com/library/ios/documentation/IDEs/Conceptual/gdb_to_lldb_transition_guide`.

REPL in Xcode

One of the more interesting functionalities available in the Xcode LLDB console is that you can run Swift REPL there. You can enter and execute Swift code when you stop the application execution in the debugger. This is very useful for debugging purposes.

> REPL can only access public types, functions, and public global variables. Local variables are not visible in REPL. If you need to work with local variables, use LLDB commands instead.

To enter an REPL console, we first have to stop the program execution and enter the LLDB debugger. There are three commands for interacting with REPL:

- **Enter REPL:** `repl`
- **Exit REPL:** `:`
- **Execute the LLDB command in REPL:** `:` `command`, for example, `:p name`

We can execute the same function as before, but now use REPL commands in the debug console, as shown here:

```
24    func LLDBTutorial() {          (lldb) repl
25        var name = "Sam"           1> :p name
26  ▶ |                              (String) $R0 = "Sam"
27    }                              1> :
                                     (lldb)
```

Now let's look at more interesting use cases of REPL. When you enter REPL, you can enter and execute Swift code. You also have access to publicly declared Swift code in your application in REPL. To summarize, you can run existing code and add new code as well.

A great use case here is adding test code directly to REPL while debugging the application. As an example, let's implement a function for skipping negative numbers in an array:

```
public func skipNegatives(a: [Int]) -> [Int] {
   return a.filter { $0 >= 0 }
}

func REPLTutorial() {
   let numbers = [2, -3, 1]
   let result = skipNegatives(numbers)
}
```

The skipNegatives function's implementation is very simple, and it is easy to check whether it works correctly in this example, but your other functions could be much bigger and harder to understand. Also, the numbers array contains only three elements and the result should contain two elements. We can easily check this by looking out for it in the debugger view.

But what if our numbers array contained 1,000 elements? It would be harder to go through the array and verify that it doesn't contain negative elements. In this example, we have 505 non-negative elements in the array:

```
func REPLTutorial() {

   let manyNumbers = makeNumbers()
   let bigResult = skipNegatives(manyNumbers)
}

public func makeNumbers() -> [Int] {
   var array = [Int]()
```

```
for _ in 0..<1000 {
  let rand = Int(arc4random_uniform(10)) - 5
  array.append(rand)
}
return array
}
```

We could write a test function in REPL to check whether all the elements are positive. Let's do that, as follows:

```
29  func REPLTutorial() {
30    let numbers = [2, -3, 1]
31    let result = skipNegatives(numbers)
32
33    let manyNumbers = makeNumbers()
34    let bigResult = skipNegatives(manyNumbers)
35
36  }
```

```
(lldb) repl
1> func isAllPositive(ar: [Int]) -> Bool {
2.    let negatives = ar.filter { $0 < 0 }
3.    return negatives.count == 0
4. }
5>
6> isAllPositive( skipNegatives([1, 2, -4, 7, 9, -1, 5, 12, -12, 24]))
$R0: Bool = true
7> isAllPositive([1, 2, -4])
$R1: Bool = false
8>
```

First, as usual, run the program and stop at a breakpoint. The next step is to enter REPL and write an `isAllPositive` function to check whether all the numbers are positive. Then, just call `skipNegatives` and `isAllPositive`, and see whether the result is `true`:

```
(lldb) repl
1> func isAllPositive(ar: [Int]) -> Bool {
2.    let negatives = ar.filter { $0 < 0 }
3.    return negatives.count == 0
4. }
5>
6> isAllPositive( skipNegatives([1, 2, -4, 7, 9, -1, 5, 12, -12, 24]))
$R0: Bool = true
7> isAllPositive([1, 2, -4])
$R1: Bool = false
8>
```

 If you are going to use a test function more than once in REPL, it is better to create a Swift source file for debugging purposes and add it there. Then, you can call it from REPL.

Console logs

The other powerful debugging tool, which you should already be familiar with, is the console log. Console logs can be used to log all types of information, including:

- Operation results
- Activities
- Performance measurement

To log a statement to the console, you can use one of these functions:

- `print`
- `debugPrint`

Both of these functions accept any type.

To provide custom text formatting for print functions, you must conform to the `CustomStringConvertible` protocol, and for `debugPrint`, the `CustomDebugStringConvertible` protocol. Both of these protocols require the implementation of only one property. Let's create a simple `Person` type and implement custom log formatting:

```
struct Person {
  let name: String
  let age: Int

extension Person:
  CustomStringConvertible, CustomDebugStringConvertible {

  // CustomStringConvertible
  var description: String {
    return "Name: \(name)"
  }

  // CustomDebugStringConvertible
  var debugDescription: String {
    return "Name: \(name) age: \(age)"
  }
}
```

Now, when `print` or `debugPrint` is called with an instance of a `Person` type, it will show a custom description.

A more interesting use case of console logs is a method's performance logging. We would like to know how much time a particular method or piece of code took to run.

The first idea would be to use NSDate to measure time. NSDate works well, but there is a better solution for this in the QuartzCore framework—a CACurrentMediaTime function. It returns the result based on mach_absolute_time. The base pseudocode for our case would be as follows:

```
let startTime = CACurrentMediaTime()
// Perform code that we need to measure.
let endTime = CACurrentMediaTime()
print("Time - \(endTime - startTime)")
```

We would like to use this performance-measuring function very often, and it would be very useful to extract this code into a separate reusable function. Because Swift supports the functional programming style, we can easily do this.

We will create a measure function. It will take another function that will perform a task for which we need to measure time:

```
func measure(call: () -> Void) {
  let startTime = CACurrentMediaTime()
  call()
  let endTime = CACurrentMediaTime()

  print("Time - \(endTime - startTime)")
}
```

Now, let's say we want to find out how much time it takes to create 1,000 instances of a Person type:

```
for i in 0...1000 {
  let person = Person(name: "Sam", age: i)
}
```

What we need to do is simply wrap this code into a closure and pass it to the measure function:

```
measure {
  for i in 1...1000 {
    let person = Person(name: "Sam", age: i)
  }
}
```

Running this measurement in the playground would give us the following result:

Time 0.000434298478008714.

Performance measuring in unit tests

When you create a new project, Xcode creates a unit test target for that project with the name `ProjectName` + `Tests`. If you are not familiar with unit testing, you can read about testing in Xcode at `https://developer.apple.com/library/ios/documentation/DeveloperTools/Conceptual/testing_with_xcode`.

Xcode will also create a simple unit test file for you. In our project, it's `Swift_ToolkitTests.swift`. The unit test has three main methods, with different purposes:

- `setup`
- `teardown`
- `test`

 The unit test function must begin with the `test` prefix, like this for example:
```
func testCreatingPerson
func testChangingName
```

The names of these functions reflect their purposes. The `setup` function performs additional setup before the unit test is run, and `teardown` performs a cleanup, but the most interesting function for us is `test`, which performs testing.

The `XCTestCase` unit test class has a `measureBlock` function that works in a way similar to the `measure` function that we have implemented. Let's implement a unit test to measure the performance for creating 1,000 people.

First, we need to make a `Person` type and other types in our application available to the unit test target. To do so, we need to import an app module with the `@testable` attribute—`@testable import {ModuleName}`. Now, all the `public` and `internal` types and methods in that module become available in the unit test target:

 To enable `@testable`, the **Enable Testability** project build setting must be set to **Yes**. Xcode sets it to **Yes** for the **Debug** mode by default. You should never enable it for release mode.

```
@testable import Swift_Toolkit

func testCreatingPeoplePerformance() {
    measureBlock() {
```

```
    for i in 1...1000 {
        _ = Person(name: "Sam", age: i)
    }
  }
}
```

When you run the unit tests, by going to **Product** | **Test** or using the *CMD + U* shortcut, you will see the performance characteristics on the right-hand side of the function name. When you click on it, you will see more details and a button for setting the baseline performance values, which will be used to compare future measurements.

The `measureBlock` runs a block of code a few times and shows an average time. You can see the performance for 10 different iterations in a detailed popup, as shown in this screenshot:

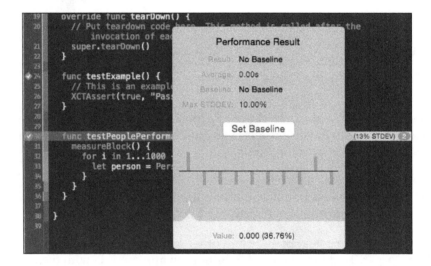

Let's now set a baseline and run the unit test again. The test passes!

The purpose of performance unit tests is to measure performance and make sure that it doesn't decrease dramatically. By default, the allowed standard deviation is 10%. This means that if the performance for code decreases by more than 10%, the test fails. Let's try to simulate this and see what happens. To simulate the extra work, we will add a delay in the `person` initializer:

```
init(name: String, age:Int) {
  self.name = name
  self.age = age
  usleep(100)
}
```

Now let's run the test again. You will see that test fails and shows a red sign next to the test function's name.

In this way, unit testing allows you to both measure performance and make sure that it doesn't decrease while you are working on the application.

Instruments

The last tool that we are going to take a look at in this chapter is instruments. Although we are mentioning it towards the end of the chapter, it's is the most powerful tool for measuring all sorts of characteristics of an application: performance, memory usage and leaks, networking, monitoring, animation, hard drive, and file activity.

The easiest way to launch Instruments for the application is by going to **Product | Profile** or by using the *CMD + I* keyboard shortcut. This will launch the instrument for the current target and show you the available instrument measurement templates. We will choose a **Time Profiler** template and click on **Record**. This will start the application and record the performance for every called function. Now we can analyze the functions' performance:

Instruments is a very powerful tool, and it would take a separate chapter to cover its functionality completely. If you are not familiar with Instruments, you should read more about it at https://developer.apple.com/library/ios/documentation/ DeveloperTools/Conceptual/InstrumentsUserGuide.

A reminder

Whenever you are doing any performance measurement, do it in release mode. The Swift compiler performs many optimization steps in the release mode and dramatically increases performance. To set release mode, go to **Product** | **Scheme** | **Edit Scheme** | **Run**, and adjust the **Build Configuration** setting to **Release**. Always use release mode for performance testing.

Summary

In this chapter, we covered many tools that would boost your productivity. REPL and Playgrounds are perfect for trying out new code and quick code prototyping. Playgrounds can also be used to create interactive documentation and tutorials. Then we covered debugging tools (such as LLDB and REPL) in Xcode, which are very useful for checking the results of operations performed at runtime. The performance of an application can be measured in Instruments or using the console log. To make sure that the performance doesn't decrease, you should use unit testing.

Basically, in this chapter, you learned many tools for discovering slow and problematic code, and in the next chapter, you will learn how to improve and optimize it.

Improving Code Performance

Achieving good code performance is an important and desirable task. Everyone would like to have applications that perform well. In this chapter we are going to cover these performance topics:

- Understanding performance optimization
- Optimization checklist
- Constants and variables
- Method calls
- Intelligent code
- Value objects and reference objects
- Swift arrays and unsafe C arrays
- Avoiding Objective-C

Understanding performance optimization

The first rule of optimization is—don't optimize. You should always remember this phrase by Donald Knut:

Premature optimization is the root of all evil

This is a very true and correct statement. You should start doing performance optimization only when you see a performance problem and you have found what is causing it.

There are two types of performance optimization:

- Explicit
- Implicit

Explicit

Explicit performance optimization is a technique that is directed at a specific slow piece of code. This type of optimization requires significant code changes that could worsen code readability. You do explicit performance optimization by changing the algorithm to a more effective one. Using more memory for the cache could also increase performance.

Implicit

Implicit performance optimization is the technique of applying language-specific, in our case Swift-specific, features that lead to better performance. Implicit code performance doesn't require significant code changes. It doesn't have any negative impact on the code readability and sometimes makes it better. I call it implicit because you can apply it everywhere in the code and it becomes invisible to you after some time.

Explicit performance optimization is a very popular and broad topic that is covered in many books about algorithms and data structures. Implicit on other hand is directly related to the Swift programming language and is a very interesting topic we are going to have a look at.

Optimization checklist

Before doing any optimization and performance measurement, you should follow these steps:

1. Enable the **Release** mode: The Swift compiler does a lot of code optimization and improves performance in the **Release** mode. To enable the **Release** mode go to **Product | Scheme | Edit Scheme | Run**, select the **Info** tab, and select **Release** in the **Build Configuration** setting.

2. Disable safety checks: Disabling safety checks could improve application performance; but as its name suggests, it affects safety and disabling is not 100-percent safe and should be applied carefully. An example of a safety check Swift does is checking array bounds before accessing the memory. If you disable safety checks Swift won't do that.

Disabling safety checks is **a Swift Compiler – Code Generation** setting
that is available in the target **Build Settings**. To disable safety checks select
Project | Build Settings, then search for the **Disable Safety Checks** setting
and set it to **Yes** for the **Release** mode.

3. Enable the **Fast, Whole Module Optimization** level: By default, the Swift
   compiler does optimization only to one file at a time. It does it as though in
   sandbox environment for every file. The optimization of one file has no effect
   on the other files.

Whole Module Optimization enables optimization for all the source files in the
module at once. All the source files are evaluated and optimized together. It is very
useful because we often declare a type in one file and use it in another. One of the
optimizations that **Whole Module Optimization** does is searching for declarations
with the `internal` type that aren't overridden anywhere and adding the `final`
declaration optimization for them.

To enable **Whole Module Optimization**, select **Optimization Level** for the Swift
compiler in **Build Setting** and select the **Fast, Whole Module Optimization [-O
-whole-module-optimization]** option for the **Release** mode.

Enabling this setting increases build time. You should use it for release builds and
performance testing. While developing and debugging, it would be better to disable
this setting to speed up the compiling time.

There are two different types of Optimization Level settings available in Xcode with
different purposes and options:

- **Apple LLVM - Code Generation**
- **Swift Compiler – Code Generation**

If you select the **Optimization Level** setting, you can see all the available options
with a detailed description in the **Quick Help** section in the **Utilities** panel.

By default in the **Release** mode Xcode uses these settings:

- **Apple LLVM, GCC_OPTIMIZATION_LEVEL: Fastest, Smallest [-Os]**
- **Swift Compiler, SWIFT_OPTIMIZATION_LEVEL: Fastest [-O]**

You can try enabling other optimization settings. For example using **Fastest, Aggressive Optimizations:[-Ofast]** could improve application performance.

Swift code compilation

The LLVM first turns your source code into pseudocode. In the next step it gets optimized and compiled into Assembly code.

You can perform these code-processing steps manually from the command line by using `swiftc` in the Swift compiler. To see all available options for the Swift compiler, open `Terminal.app` and execute the `--help` command:

```
xcrun swiftc --help
```

You will see the available compilation modes and options. The ones we are looking for are:

- `-emit-assembly`
- `-emit-ir`
- `-emit-silgen`
- `-emit-sil`

These compilation modes allow you to apply different compilation steps to the Swift source file. As an example, we could emit `sourceFile.swift` into the canonical SIL representation and write the result to the `outputFile` with this command:

```
swiftc -emit-sil sourceFile.swift -o outputFile
```

We will cover the compilation process in greater detail in *Chapter 8, Discovering All the Underlying Swift Power*.

Constants and variables

Using constant has an impact on code readability. It makes code clearer and safer. Using constants instead of variables could also have performance benefits. When you use a constant you give the compiler a clear hint that this value won't be changed. The Swift compiler can apply inline optimization to use a value of that constant and not allocate memory for it.

In simple examples, the Swift compiler could do the same optimization for variables as well. Let's analyze the result for this simple example of iterating and calculating a sum. The performance is the same for variables and constants in this example.

```
var result = 0
for _ in 0...10000000 {
  let a = Int(arc4random())
  result += a
}
// Average Time - 0.162666518447804

var result = 0
for _ in 0...10000000 {
  var a = Int(arc4random())
  result += a
}
// Average Time - 0.160957522349781
```

If we look at a more complex example, we will see that constants perform the same as, or even better than, variables. It might seem as if a version using variables should perform faster, because there is no need to allocate memory for new constants on every operation, but the Swift compiler is smart enough to perform intelligent optimization so that they behave the same.

```
var result = 0
for _ in 0...100000000 {
  let a = Int(arc4random_uniform(10))
  let b = a + Int(arc4random_uniform(10))
  let c = b * Int(arc4random_uniform(10))

  result += c
}
// Average Time - 12.6813167635002

var result = 0
for _ in 0...100000000 {
  var a = Int(arc4random_uniform(10))
  a += Int(arc4random_uniform(10))
  a *= Int(arc4random_uniform(10))
  result += a
}
// Average Time - 12.6813167635102
```

So the general advice is: Prefer using constants. They make the code safer and clearer and also have a positive impact on performance. Variables in some situations could also improve code readability, as in the example earlier where we had to do some math calculations and changing the value in the variable actually made the code clearer.

Constants are so much better than variables that Xcode shows a warning when it detects a variable that was never mutated and suggests you change it to a constant.

```
23        for _ in 0...10000000 {
24            let a = Int(arc4random())
25            result += a
```
⚠ Variable 'a' was never mutated; consider changing to 'let' constant
Fix-it Replace "var" with "let"
```
29    }
```

Method calls

Before discussing Swift method calls optimization, it would be very useful to have a look at different types of method call implementation.

There are two main types of method call:

- **Static**: Static method binding means that, when you call a method on the object, the compiler knows that you are calling exactly this method on exactly this class. C is an example of a language with static method binding.

- **Dynamic**: On other hand, dynamic has a weak binding between the method and the object. When you call a method on the object there is no guarantee that an object can handle this method call. Objective-C has a dynamic method binding. That's why you can see the `object does not respond to selector` error in Objective-C.

Objective-C is a dynamic-type language and it has a dynamic runtime. Calling a method is called **message sending**. You send a message to the target.

```
[dog bark] // dog is a target and bark is a message
```

This looks like a normal method call, but after compilation it would actually look like:

```
objc_msgSend(dog, @selector(bark))
```

Objective-C uses dynamic method binding. It means that the message and the receiver are stored separately. When you send a `bark` message to the `dog` object, the dog class has to look up if it has a bark method and if it can handle it. This process is called dynamic method binding. The implementation would look like this:

```
id objc_msgSend ( id obj, SEL _cmd, ... )
{
    Class c = object_getClass(obj);
    IMP imp = CacheLookup(c, _cmd);
    if (!imp) {
        imp = class_getMethodImplementation(c, _cmd);
    }
    jump imp(obj, op, ...);
}
```

Swift uses a static method binding. It uses a vtable — Virtual Method Table — for storing methods. Vtable is an array of function pointers. That means that a class has a list of its methods with the memory address of that method implementation. When you call a method in Swift, you are calling it on the specific type. The binding between the method and the object you are calling this method on is very strong and done at compile time.

Let's have a look how the same code would behave in Swift:

```
dog.bark()
```

Because Swift knows that you want to call a bark method on the `Dog` class, it doesn't need to do any extra lookup for the method information. It will get the function address and call it:

```
methodImplementation = dog->class.vtable[indexOfBark]
methodImplementation()
```

Swift can do even more complex optimization to method calls. If the method is not overridden, it means that call to the `bark` method will always resolve to the same function call. The Swift compiler can skip the function lookup in the vtable and inline direct function call:

```
_TFC12methodsCalls3Dog4barkfS0_FT_T_()
//this method is equal to- methodsCalls.Dog.bark()
```

> This is the mangled name of the `bark` method. We will learn more about this in *Chapter 8, Discovering All the Underlying Swift Power.*

The `_TFC12methodsCalls3Dog4barkfS0_FT_T_()` direct function call in the assembly code is translated into a simple command. Here is how the assembly pseudo code looks:

```
rbx = __TFC11Performance3DogCfMS0_FT_S0_(); // Create dog instance
r15 = *(*rbx + 0x48); //get the location of bark method
(r15)(rbx); // call the method
```

Let's compare the performance of Swift static method calls with the Objective-C dynamic method call to the see the performance difference. Let's make a simple `Number` class with an `add` method that will add two numbers (a Swift solution):

```
class Number {

  func add(x: Int, y: Int) -> Int {
    return x + y
  }
}
```

For time measurement we are using our `measure` function from the previous chapter:

```
let number = Number()
measure("Sum", times: 20) {
  var result: Int = 0
  for i in 0...1000000000 {
    result += number.add(i, y: i + 1)
  }
  print(result)
}
```

The result: `Average Time - 1.45391867654989`.

Let's see the Objective-C solution:

```
//   KKNumber.h
@import Foundation;

@interface KKNumber : NSObject

-  (NSInteger)add:(NSInteger)num number:(NSInteger)num2;

@end

//   KKNumber.m
#import "KKNumber.h"
```

```
@implementation KKNumber

- (NSInteger)add:(NSInteger)num number:(NSInteger)num2 {
    return num + num2;
}

@end

KKNumber *number = [[KKNumber alloc] init];

[Measure measure:20 call:^{
    NSInteger result = 0;
    for (int i  = 0; i < 1000000000; ++i) {
        result += [number add:i number:i + 1];
    }
    NSLog(@"Result %ld", (long)result);
}];
```

The result: `Average Time - 2.974986.`

As you can see, even a very simple function call is twice as fast in Swift. Now you know the details of Swift method and functions call implementation, it's time to jump to more practical examples.

Functions and methods

You can make code reusable by making a function or a method in four different ways:

- Global functions
- Type methods
- Static and final methods
- Instance methods

Global functions

Global functions are the simplest ones. They cannot be overridden and changed. Global functions are stored as named pointers in memory. When you call a global function it is translated to the direct memory call without any lookup in the vtable. This should be the fastest way. The assembly code for calling a global function is:

```
call        __TZFC4test3Dog5speakfMS0_FT_T_
```

Type methods

Type methods operate on the type and not instances of that type. Class methods are stored in the vtable for that class. The class method can be overridden by subclass. Because class methods can be overridden, the Swift compiler sometimes can't optimize class method calls to a direct function call as for global functions. For a better understanding of why, let's have a look at this simple example of an overridden class method:

```
class Dog {
  class func bark() {
    print("Bark")
  }
}

class BigDog: Dog {
  override class func bark() {
    print("big loud BARK")
  }
}

func getDog() -> Dog.Type {
  return arc4random() % 2 == 0 ? Dog.self : BigDog.self
}

let dog = getDog()
dog.bark()
```

We have made two simple classes: `Dog` and `BigDog`. The function `getDog` returns a `Dog.Type` class type, but it can also return `BigDog.Type`. The `dog` variable can be either a `Dog.Type` or a `BigDog.Type`. Because of that, the Swift compiler can't do direct function calls inline. It has to do a lookup for the function pointer in the vtable and that is a very cheap operation. The pseudo assembly code for this would look like:

```
rax = __TF4test6getDogFT_MCS_3Dog(); // call getDog()
*__Tv4test3dogMCS_3Dog = rax;        // convert result to Dog.Type.
(*(rax + 0x48))(rax);                // call bark method, vtable lookup
```

The Swift compiler can do direct function calls inline for overridden methods when you specify a type explicitly. In this example we call the `bark` method on the `Dog` class and the Swift compiler skips the vtable lookup:

```
Dog.bark()

// Pseudo assembly code
__TTSf4d___TZFC4test3Dog4barkfMS0_FT_T_
```

Static methods

You can declare a static type method in classes, structures and enumerations. In classes declaring a type method, using the `static` keyword is the same as using the `final class` keywords. Those two method declarations are equivalents:

```
static func speak() {}
final class func speak() {}
```

Static methods can't be overridden in the subclasses. Because they can't be overridden they don't need to be stored in the vtable. The implementation details of a static method are very similar to a global function. In the assembly code it will be translated to a direct function call, the same as for global functions. Let's add a static function to our dog class and explore how it translates to the assembly code:

```
class Dog {

  static func speak() {
    print("I don't speak")
  }
}

Dog.speak()
BigDog.speak()
```

Both calls to the `speak` method in `Dog` and the `BigDog` class are translated to a one-line assembly instruction.

```
call          __TZFC4test3Dog5speakfMS0_FT_T_
```

Instance methods

The main difference between type methods and instance methods is that instance methods can operate with instance constants and variables. Instance methods can be overridden and they need to be stored in the vtable. Let's add a name variable to our `Dog` class and a `changeName` instance method:

```
class Dog {
  var name = ""

  func changeName(name: String) {
    self.name = name
  }
}

let someDog = Dog()
someDog.changeName("Cocoa")
```

The `changeName` method will be translated to this assembly code. Get the method address from the vtable and call it with passing parameters:

```
rbx = __TFC4test3DogCfMS0_FT_S0_(); // Create Dog()
*__Tv4test7someDogCS_3Dog = rbx;  //Assign Dog instance to a
someDog variable
r15 = *(*rbx + 0x68);      // Get changeName method, vtable lookup
(r15)("Coca", 0x4, 0x0, rbx); // call method and pass arguments
```

Comparing function speed

Now you know how functions and methods are implemented and how they work. Let's compare the performance speed of those global function and the different method types. For the test we will use a simple `add` function. We will implement it as a global function, static, class type, and instance and override them in the subclass:

```
func add(x: Int, y: Int) -> Int {
   return x + y
}

class NumOperation {

   func addI(x: Int, y: Int) -> Int
   class func addC(x: Int, y: Int) -> Int
   static func addS(x: Int, y: Int) -> Int
}

class BigNumOperation: NumOperation {

   override func addI(x: Int, y: Int) -> Int
   override class func addC(x: Int, y: Int) -> Int
}
```

For the measurement and code analysis, we use a simple loop where we call those different methods:

```
measure("addC") {
   var result = 0
   for i in 0...2000000000 {
      result += NumOperation.addC(i, y: i + 1)
      // result += test different method
   }
   print(result)
}
```

Results:

All these methods perform in exactly the same way. Furthermore, their assembly code looks exactly the same, except the name of the function call:

- **Global function**: `add(10, y: 11)`
- **Static**: `NumOperation.addS(10, y: 11)`
- **Class**: `NumOperation.addC(10, y: 11)`
- **Subclass Static**: `BigNumOperation.addS(10, y: 11)`
- **Subclass overridden class**: `BigNumOperation.addC(10, y: 11)`

The assembly pseudocode for those functions looks likes this:

```
r14 = 0x0;
do {
   rbx = "Function name Here"(r14 + 0x1, r14) + rbx;
   r14 = r14 + 0x1;
} while (r14 != 0x3ea);
```

Even though the `BigNumOperation` addC class function overrides the `NumOperation` addC function when you call it directly, there is no need for a vtable lookup.

The instance method call looks a bit different.

- **Instance**:
  ```
  let num = NumOperation()
  num.addI(10, y: 11)
  ```

- **Subclass overridden instance**:
  ```
  let bigNum = BigNumOperation()
  bigNum.addI()
  ```

The one difference is that they need to initialize a class and create an instance of the object. In our example this is not such an expensive operation because we do it outside the loop and it happens only once:

```
if (rax == 0x0) {
   rax = _swift_getInitializedObjCClass (
     objc_class__TtC4test12NumOperation);
   *__TMLC4test12NumOperation = rax;
}
var_78 = _swift_allocObject(rax, 0x10, 0x7);
```

The loop with the calling instance method looks exactly the same so we will not list it again.

As you have seen there is almost no difference between global functions and static and class methods. Instance methods look a bit different but it doesn't have a big impact on performance. Though this is true for simple use cases, there is a difference between them in more complex examples. Let's have a look at this one:

```
let baseNumType = arc4random_uniform(2) == 1 ?
  BigNumOperation.self : NumOperation.self

for i in 0...loopCount {
  result += baseNumType.addC(i, y: i + 1)
}
print(result)
```

The only difference here is that, instead of specifying the NumOperation class type at compile time, we randomly return it at runtime. And because of this, the Swift compiler doesn't know which method should be called—BigNumOperation.addC or NumOperation.addC—at compile time. This small change has an impact on the generated assembly code and the performance.

Functions and methods usage summary

Global functions are the simplest ones and provide the best performance. Having too many global functions makes code hard to read and follow.

Static type methods that can't be overridden have the same performance as global functions but they also provide a namespace (type name), so our code looks clearer and without any loss in performance.

Class methods that can be overridden could lead to a performance loss and should be used when you need class inheritance. In other cases, static methods are preferred.

Instance methods operate on the instance of the object. Use instance methods when you need to operate on the data of that instance.

Make methods final when you don't need to override them. This tells the compiler that optimization and performance could be increased because of that.

Intelligent code

Because Swift is a static and strongly typed language it can read, understand, and optimize code very well. Swift tries to remove the execution of all unnecessary code. For a better explanation let's have a look at a simple example:

```
class Object {

  func nothing() {
  }
}

let object = Object()
object.nothing()
object.nothing()
```

We create an instance of the `Object` class and call a `nothing` method. The `nothing` method is empty and calling it does nothing. The Swift compiler understands that and removes those method calls. After this we have only one line of code:

```
let object = Object()
```

The Swift compiler can also obviate the creation of objects that are never used. It reduces memory usage and unnecessary function calls, which also reduces CPU usage. In our example the `object` instance is not used after removing the `nothing` method call and the creation of `Object` can be dispensed with as well. This way, Swift removes all three lines of code and we end up with no code to execute at all.

Objective-C, can't do this optimization. Because it has a dynamic runtime, the `nothing` method implementation could be changed to do some work at runtime. That's why Objective-C can't remove empty method calls.

This optimization might not seem to amount to much but let's have a look at another, slightly more complex example that uses more memory:

```
class Object {
  let x: Int
  let y: Int
  let z: Int

  init(x: Int) {
    self.x = x
    self.y = x * 2
    self.z = y * 2
  }

  func nothing() {
  }
}
```

We have added some `Int` data to our `Object` class to increase memory usage. Now the `Object` instance uses at least 24 bytes (3 * `Int` size; `Int` uses four bytes in 64-bit architecture). Let's also try to increases CPU usage by adding more instructions via a loop:

```
for i in 0...1_000_000 {
  let object = Object(x: i)
  object.nothing()
  object.nothing()
}
print("Done")
```

 Integer literals can use underscores (_) to improve readability. 1_000_000_000 is the same as 1000000000

Now we have three million instructions and we use 24 million bytes, about 24 MB. This is quite a lot for the type of operation that actually doesn't do anything. As you can see, we don't use the result of the loop body. For the loop body Swift does the same optimization as in the previous example and we end up with an empty loop:

```
for i in 0...1_000_000 {
}
```

The empty loop can be skipped as well. As a result, we have saved 24 MB of memory usage and three million method calls.

Dangerous functions

There are some functions and instructions that sometimes don't provide any value for the application but the Swift compiler can't skip them and they could have a very negative impact on performance.

Console print

Printing a statement to the console is usually used for debug purposes. The `print` and `debugPrint` instructions aren't removed from the application in the Release mode. Let's explore this code:

```
for i in 0...1_000_000 {
  print(i)
}
```

The Swift compiler treats `print` and `debugPrint` as a valid and important instruction that can't be skipped. Even though this code does nothing, it can't be optimized because Swift doesn't remove the `print` statement. And as a result we have one million unnecessary instructions. The assembly code for this is:

```
mov      qword [ss:rbp+var_20], rbx
inc      rbx                  //increase i
mov      rdi, r14             // save stack state for function call
mov      rsi, r15
call     __TFSs5printurFq_T_  //call print
cmp      rbx, 0xf4241         // check loop condition i > 1_000_000
jne      0x100155fb0          // continue loop if condition is true
```

As you can see even very simple code that uses the `print` statement can decrease application performance very dramatically. The loop with the 1_000_000 `print` statement takes five seconds and that's a lot. It's even worse if you run it in Xcode; it will take up to 50 seconds.

It gets even worse if you add a `print` instruction to the `nothing` method of an `Object` class from the previous example:

```
func nothing() {
    print(x + y + z)
}
```

In that case, a loop where we create an instance of `Object` and call `nothing` can't be eliminated because of the `print` instruction. Even though Swift can't eliminate execution of that code completely it does the optimization by removing the creation instance of `Object` and calling the `nothing` method, and turns it into a simple loop operation. The compiled code after optimization will look like this:

```
// Initial Source Code
for i in 0...1_000 {
    let object = Object(x: i)
    object.nothing()
    object.nothing()
}

// Optimized Code
var x = 0, y = 0, z = 0
for i in 0...1_000_000 {

    x = i
    y = x * 2
    z = y * 2
```

```
    print(x + y + z)
    print(x + y + z)
}
```

As you can see, this code is far from perfect and supplies a lot of instructions that actually don't give us any value. There is a way to improve this code so the Swift compiler would perform the optimal code optimisation.

Removing print logs

To solve this performance problem we have to remove the `print` statements from the code before compiling it. There are a few ways to do that.

Comment out

The first idea is to comment out all `print` statements in the code in the Release mode.

```
//print("A")
```

This would work but the next time you want to enable logs, you would need to uncomment that code. This is a very bad and painful practice. There is a better solution.

> Commented code is bad practice in general. You should be using a source code version control system, such as Git, instead. This way you can safely remove unnecessary code and find it in history if you need it someday.

Using build configurations

We can enable `print` only in the **Debug** mode. To do this, we will use a build configuration to conditionally exclude some code. First we need to add a Swift compiler custom flag. To do that:

Select a Project target— **Build Settings – Other Swift flags** setting in the **Swift Compiler – Custom Flags** section and add the **–D DEBUG** flag for the **Debug** mode:

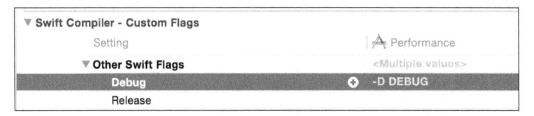

After this you can use the DEBUG configuration flag to enable code only in the Debug mode. We will define our own print function that will generate print statement only in debug mode. In the Release mode, that function would be empty and the Swift compiler will successfully eliminate it:

```
func D_print(items: Any..., separator: String = " ", terminator:
  String = "\n") {
  #if DEBUG
    print(items, separator: separator, terminator: terminator)
  #endif
}
```

Everywhere instead of print we will now use D_print:

```
func nothing() {
  D_print(x + y + z)
}
```

 You can also create a similar D_debugPrint function.

Swift is very smart and does a lot of optimization but we also have to make code clear for people to read and for the compiler to optimize.

 Using a preprocessor adds complexity to the code. Use this wisely and only in those situations when normal if conditions won't work, like in our D_print example.

Using nonoptimizable constants

Some types can't be optimized as well as others and using constants of that type can't prevent the code from being eliminated by the Swift compiler.

Let's have a look at this simple example:

```
class Optimizable {
  let x = 10
}

// Use-case
let o = Optimizable()
```

The Swift compiler can eliminate this code. Let's have a look at a more complex example:

```
class Optimizable {
   let x = 10
   let a = ""
}

// Use-case
let o = Optimizable()
```

This code can't be completely eliminated by the Swift compiler. By adding a very simple `String` constant, we have added much more complexity to the source-code. To understand why it happened we need to explore the assembly code:

```
if (*__TMLC4test11Optimizable == 0x0) {
    *__TMLC4test11Optimizable = _swift_getInitializedObjCClass();
    // Initialize objc_class__TtC4test11Optimizable
}
rax = _swift_allocObject();
*(rax + 0x10) = 0xa;
*(rax + 0x18) = "";
```

From the name of the `getInitializedObjCClass` we could make an assumption that this method does some Objective-C class type initialization. It might seem very strange because we haven't used any Objective-C types in our code. We have added a simple empty string constant: `let a = ""`.

The thing is that the Swift `String` provides seamless interoperability with the Objective-C `NSString` type. Because of this, when we use a Swift `String` it allocates some additional data to perform bridging to `NSString`. Here is how that metadata `objc_class__TtC4test11Optimizable` looks:

```
objc_class__TtC4test11Optimizable:
dq          __TMmC4test11Optimizable; metaclass,
dq          _OBJC_CLASS_$_SwiftObject; superclass
dq          __objc_empty_cache; cache
dq          __objc_empty_vtable; vtable
dq          0x1001bf7e1; data (Swift class)
```

There are more types that can't simply be eliminated by the Swift compiler when they are used with simple constants:

- String
- Array

- Custom Class objects
- Closures
- Set
- Dictionary

```
class NotOptimizableTypes {

    let a: String = ""
    let b: String? = nil
    let c: Array<Int> = [1]
    let obj = Object()
    let d: Int -> Int = { $0 + 1 }

    let e: Set<Int> = [1]
    let f: Dictionary<Int, Int> = [1 : 1]
}
```

The interesting behavior happens if we try to use those types in a structure instead of a class. We see different behavior because Swift structures aren't exposed for use in Objective-C. That's why the Swift compiler can eliminate many of them: String, Array, Class, and Closures. Set and Dictionary aren't eliminated even if they are used in the structure.

```
struct NotOptimizableInStruct {

  let a: String = ""
  let b: Array<Int> = [1]
  let obj = Object()
  let c: Int -> Int = { $0 + 1 }
}
```

Moving those constant to the initializer doesn't solve the problem.

The solution to that issue is that you shouldn't use any constants that aren't used and don't provide any value for the application.

Improving speed

There are a few techniques that can simply improve code performance. Let's jump directly to the first one.

Final

You can make a function and property declaration with the `final` attribute. Adding the `final` attribute makes it nonoverridable. Subclasses can't override that method or property. When you make a method nonoverridable there is no need to store it in the vtable and the call to that function can be performed directly without any function address lookup in the vtable:

```
class Animal {

  final var name: String  = ""
  final func feed() {
  }
}
```

As you have seen, the `final` method performs faster than nonfinal. Even so small an optimization can improve application performance. It not only improves performance but also makes the code more secure. This way you disable the method from being overridden and prevent unexpected and incorrect behavior.

Enabling the **Whole Module Optimization** setting achieves very similar optimization results but it's better to mark a function and property declaration explicitly as `final`: This reduces compiler work and speeds up compilation time. The compilation time for big projects with **Whole Module Optimization** can take up to five minutes in Xcode 7 Beta 6.

Inline functions

As you have seen Swift can do optimization and make some function calls inline. This way there is no performance penalty for calling a function. You can manually enable or disable inline functions with the `@inline` attribute:

```
@inline(__always) func someFunc () {
}

@inline(never) func someFunc () {
}
```

Even though you can manually control inline functions it's usually better to leave it to the Swift compiler to do that. Depending on different optimization settings, the Swift compiler applies a different inlining technique.

The use case for `@inline(__always)` is a very simple one-line function that you want always to be inlined.

Value objects and reference objects

In the previous chapter, you learned the benefits of using immutable value objects. Value objects not only make code safer and clearer, they also make it faster. Value objects have better speed performance than reference objects and here is why. As an example of a value object, we will use structures in this chapter.

Memory allocation

Value objects can be allocated on the stack memory instead of the heap memory. Reference objects need to be allocated on the heap memory because they can be shared between many owners. Because value objects have only one owner they can be safely allocated on the stack. Stack memory is way faster that heap memory.

The second advantage is that value objects don't need reference counting memory management. As they can have only one owner, there is no such thing as reference counting for value objects. With **ARC (Automatic Reference Counting)**, we don't need to think so much about memory management and it mostly looks transparent for us. Even though using reference-object and value-object code looks the same, ARC adds extra retain and release method calls for reference objects. Let's look at a very simple example of a structure and class that represent a number:

```
struct NumberValue {
  let x: Int
}

class NumberReference {
  let x: Int
  init(x: Int) {
    self.x = x
  }
}
```

As an example we will write exactly the same code using `NumberValue` and `NumberReference` and compare the generated assembly code:

```
var x = NumberValue(x: 1)
var xres = x.x
x = NumberValue(x: 2)
xres += x.x

var y = NumberReference(x: 10)
var yres = y.x
y = NumberReference(x: 20)
yres += y.x
```

The two lines of code for creating and using the `NumberValue` structure look very simple. In the assembly, it has three lines of code that do the following:

- Create the `NumberValue` object
- Assign it to x variables
- Save x number to the `xres` variable

```
rax = __TFV4test11NumberValueCfMS0_FT1xSi_S0_(0x1);
*__Tv4test1xVS_11NumberValue = rax;
*__Tv4test4xresSi = rax;

// NumberValue(x: 2)
rax = __TFV4test11NumberValueCfMS0_FT1xSi_S0_(0x2);
*__Tv4test1xVS_11NumberValue = rax;
*__Tv4test4xresSi = *__Tv4test4xresSi + rax;
```

As you can see the code for creating the first number object and the second one looks exactly the same. Now let's have a look the assembly code for the `NumberReference` objects:

```
rax = __TFC4test15NumberReferenceCfMS0_FT1xSi_S0_(0xa);
*__Tv4test1yCS_15NumberReference = rax;
*__Tv4test4yresSi = *(rax + 0x10);
```

As you can see the first three lines look almost the same. It creates an instance of `NumberReference`, assigns it to the variable, gets the x number, and saves it to the `yres` variable. The code for creating second instance is more interesting:

```
// NumberReference(x: 10)
rax = __TFC4test15NumberReferenceCfMS0_FT1xSi_S0_(0x14);
rdi = *__Tv4test1yCS_15NumberReference;
*__Tv4test1yCS_15NumberReference = rax;
_swift_release(rdi, r14);
rax = *__Tv4test1yCS_15NumberReference;
*__Tv4test4yresSi = *__Tv4test4yresSi + *(rax + 0x10);
```

As you can see it has three lines more than `NumberValue`. We have assigned a new instance to the y variable; the old `NumberReference` got out of a scope and needs to be released. Those three lines are related to the `_swift_release` function. If you further analyze the assembly code for working with reference objects, you will also discover another ARC function: `_swift_retain;`.

Now you know the main performance difference between value types and reference types, let's see how they perform. For that, let's use the number type and perform some calculations in the loop.

```
var result = NumberValue(x: 0)
for i in 0...1_000 {
  var x = NumberValue(x: result.x + i)
  result = x
}

print(result)

var refResult = NumberReference(x: 0)
for i in 0...1_000 {
  var x = NumberReference(x: refResult.x + i)
  refResult = x
}
print(refResult)
```

Output:

NumberValue: 500500

NumberReference: 500500

The first loop with the `NumberValue` structure is completely eliminated by the Swift compiler at compile time. The calculating loop is turned into a simple integer result; 500500 in hexadecimal equals 0x7a314. Here is the assembly pseudocode for the first loop:

```
var_30 = 0x7a314;                              // save 500500
__TFSs5printU__FQ_T_(var_30, 0x1001ba538); // call print
```

As you can see there no loop execution, the result is evaluated at the compile time.

The second loop using the `NumberReference` reference objects can't be eliminated at compile time. The assembly pseudocode structure looks exactly the same as the source code:

```
if (r14 == 0x0) {
  r14 = _swift_getInitializedObjCClass();
  *__TMLC4test15NumberReference = r14;
}
r15 = _swift_allocObject();
*(r15 + 0x10) = 0x0;
rbx = 0x0;
do {
```

```
    r13 = rbx + 0x1;
    rbx = rbx + *(r15 + 0x10);
    r12 = _swift_allocObject();
    *(r12 + 0x10) = rbx;
    _swift_release(r15, 0x18);
    r15 = r12;
    rbx = r13;
} while (r13 != 0x3e9);
var_38 = r12;
__TFSs5printU__FQ_T_(var_38, r14);
```

As you see, using the value object gives a much bigger performance win. As an example, let's measure the performance for that operation but increase the loop iterations to 1_00_000_000:

```
NumberValue Time - 0.000438838000263786
NumberReference Time - 8.49874957299835
```

This is not really a fair performance measurement because the variant with the value object actually doesn't do any execution. To compare the actual execution speed let's run this code in debug mode:

 You shouldn't measure performance in the Debug mode.

The results are:

NumberValue Time - 4.31753185200068

NumberReference Time - 15.4483174900015

The difference is still impressive; Value objects perform up to four times faster

Swift arrays and unsafe C arrays

Everyone knows that C is a super-fast programming language and when you hit a performance problem people go to C for help. In Objective-C, it is very easy to use C functions and types; its name says it all—C with Objects.

Swift also has support for interacting with C types and pointers. Even though it's available, it's considered as a dangerous operation because you need to do memory management manually. You need to allocate and destroy memory. Those types are called **Unsafe** in Swift and start with the Unsafe prefix — for example:

- UnsafePointer
- UnsafeMutablePointer
- UnsafeBufferPointer

 Avoid using C pointers in Swift. It adds a lot of complexity to the code.

There are three main use scenarios for UnsafePointers:

- Function parameters
- Creating a pointer to an existing variable
- Allocating memory for the pointer

Function parameters

First let's learn how to use pointers. When designing your API in Swift, you shouldn't use UnsafePointers, but you can find a situation when you need to interact with the C API: a Core Foundation, for example. C pointers would be exposed to Swift as:

- const Int * as UnsafePointer<Int>
- Int * as UnsafeMutablePointer<Int>

When you call a function in Swift with the UnsafePointer parameter, you can pass a variable of the same type as an in-out argument, by using the & sign or an array:

```
var num = 10
var ar = [1, 2]

func printNumber(x: UnsafePointer<Int>) {
   x.memory
}

printNumber(&num)
printNumber(ar)
```

It is also possible to pass `nil`, but in that case our function would have a null pointer as an argument and accessing its memory would crash the application:

```
printNumber(nil)
```

You will see the following on the screen:

Execution was interrupted, reason: EXC_BAD_ACCESS (code=1, address=0x0).

Using `UnsafeMutablePointer` is very similar. The main difference between `UnsafeMutablePointer` and `UnsafePointer` is that a mutable pointer can mutate the value of the variable it points to. When using arrays as an argument for a function with the `UnsafeMutablePointer` parameter, they also need to be passed as in-out parameters.

```
func changeNumber(x: UnsafeMutablePointer<Int>) {
  x.memory = 9901
}

changeNumber(&num)
changeNumber(&ar)
num // 9901
ar // [9901, 2]
```

Creating a pointer to an existing variable

When you create a pointer you can connect it to an existing variable with the `initialize` method. The `initialize` method would return a function that could be used to set a new value for that variable.

```
var num = 10
var ar = [1, 2]

var numPtr = UnsafeMutablePointer<Int>.initialize(&num)
numPtr(10)
num //10

var numArPtr = UnsafeMutablePointer<[Int]>.initialize(&ar)
numArPtr([1])
ar  //[1]
```

Allocating memory for the pointer

The other way to work with pointers is to allocate a memory for them. The `alloc` method has one parameter: the number of objects it will allocate memory for. After allocating a memory for the pointer you can use it. Finally, you need to deallocate the memory used by the pointer.

```
var numberPtr = UnsafeMutablePointer<Int>.alloc(1)
numberPtr.memory = 20
numberPtr.memory // 20
numberPtr.dealloc(1)
```

`UnsafeMutablePointer` has many useful methods that you can use, such as the `successor` and `predecessor` methods for moving pointers forward and backward, `subscript` for accessing random pointer indices, and many others.

You can read more about interacting with C pointers at `https://developer.apple.com/library/ios/documentation/Swift/Conceptual/BuildingCocoaApps/InteractingWithCAPIs.html`.

Now you have learned how to work with C pointers in Swift let's jump to our main goal: measuring how fast it would be to work with pointer arrays.

Comparing Swift arrays with unsafe C arrays

For comparison, we will make an array of random numbers and sort them. The main goal is not to find the most efficient way to sort numbers in the array but to compare performance when working with C `UnsafePointers` and Swift array types.

First let's make a C-style variant with `UnsafeMutablePointer`:

```
let count = 3_000_0
measure("C Arrays") {
  let array = UnsafeMutablePointer<Int>.alloc(count)
  for a in 0..<count {
    array[a] = Int(arc4random())
  }

  // Sort
  for _ in 0..<count {
    for j in 0..<count - 1 {
      if array[j] > array[j + 1] {
        swap(&array[j], &array[j + 1])
      }
    }
  }
  array.dealloc(count)
}
```

The **result is:** `Average Time - 1.31680929350041.`

Now let's make the same solution by using Swift arrays:

```
let count = 3_000_0
measure("Swift Arrays") {
  var array = Array(count: count, repeatedValue: 0)

  for i in 0..<count {
    array[i] = Int(arc4random())
  }

  // Sort
  for _ in 0..<count {
    for j in 0..<count - 1 {
      if array[j] > array[j + 1] {
        swap(&array[j], &array[j + 1])
      }
    }
  }
}
```

The **result is:** `Average Time - 1.30709397329978.`

The Swift arrays have the same performance as when working with `UnsafePointers`.

As you can see the code looks very similar. The initialization of arrays and sort algorithms looks exactly the same in both variants. It's because both `Array` and `UnsafeMutablePointer` have a subscript method. The only difference is in the way we create arrays:

For `UnsafeMutablePointer`:

```
let array = UnsafeMutablePointer<Int>.alloc(count)
...
array.dealloc(count)
```

For Swift array:

```
    var array = Array(count: count, repeatedValue: 0)
```

In general, Swift Array provides more functionality and it's much easier to work with. As an example, Array has sort, filter and many other methods but `UnsafeMutablePointer` doesn't.

A summary of working with pointers

To keep it short— Swift arrays are preferred, and here is why.

Working with pointers is an unsafe and dangerous operation. You need to manually allocate and release memory. Accessing memory with pointers is also very dangerous because you can access other memory that doesn't belong to you.

UnsafePointers and Swift arrays have the same performance characteristics.

Avoiding Objective-C

You have learned that Objective-C (with its dynamic runtime) in most cases performs more slowly than Swift. Interoperability between Swift and Objective-C is done so seamlessly that sometimes we can use Objective-C types and its runtime in the Swift code without knowing that.

When you use Objective-C types in Swift code, Swift is actually using the Objective-C runtime for method dispatch. Because of that, Swift can't do the same optimization as for pure Swift types. Let's have a look at a simple example:

```
for _ in 0...100 {
    _ = NSObject()
}
```

Let's read this code and make some assumptions about how the Swift compiler would optimize that code. The NSObject instance is never used in the loop body, so we could eliminate creating an object. After that we would have an empty loop that could also be eliminated. So we would remove all the code from execution.

Let's see what is happening in reality by looking at generated assembly pseudocode:

```
rbx = 0x65;
  do {
     rax = [_OBJC_CLASS_$_NSObject allocWithZone:0x0];
     rax = [rax init];
     [rax release];
     rbx = rbx - 0x1;
     COND = rbx != 0x0;
  } while (COND);
```

As you can see no code gets eliminated. This is happening because Objective-C types use the dynamic runtime dispatch method, called message sending.

All the standard Frameworks such as Foundation and UIKit are written in Objective-C and all the types—such as NSDate, NSURL, UIView, and UITableView—use the Objective-C runtime. They don't perform as fast as Swift types but we get all those frameworks available for use in Swift and this is great.

There is no way to remove the Objective-C dynamic runtime dispatch from the Objective-C types in Swift, so the only thing we can do is to learn how to use them wisely.

Avoiding exposing Swift to Objective-C

We can't remove runtime behavior from Objective-C types but we can stop Swift types from using the Objective-C runtime.

When a Swift class inherits from an Objective-C class it also inherits its dynamic runtime behavior. This also makes it available for usage in Objective-C code. Because it inherits its dynamic behavior the Swift compiler can't perform optimal optimization (as shown in the earlier example with NSObject inside a loop). Let's make a simple class that inherits from Objective-C and explore its behavior:

```
class MyNSObject: NSObject {
}

for _ in 0...100 {
    _ = MyNSObject()
}
```

This code can't be eliminated and its assembly code looks very similar to that shown in the earlier example. We could improve that behavior very simply by removing the NSObject inheritance. We can do that in this example because we don't use any features from NSObject.

```
class MyObject {
}

for _ in 0...100 {
    _ = MyObject()
}
```

In that case, the Swift compiler is able to perform optimal optimization and eliminates all the code from execution. It removes the creation of the MyObject inside the loop and eliminates the empty loop afterwards.

As you have seen, using Objective-C classes in Swift makes the Swift compiler less powerful. Inherit and use Objective-C classes only if it's required in these cases:

- Expose Swift types to Objective-C
- Need to subclass from Objective-C class, such as UIView, UIViewController and so on. Subclass only when it's really required

Dynamic

There is one more dangerous attribute that adds Objective-C dynamic runtime behavior to your type: the `dynamic` keyword. When you make a member declaration with the `dynamic` modifier it adds the Objective-C runtime to the class. The access to that member will never be statically inlined and will be always dispatched dynamically by using the Objective-C Target-Action mechanism. Let's examine this simple example:

```
class MyObject {
  dynamic func getName() -> String {
    return "Name"
  }

  dynamic var fullName: String {
    return "Full Name"
  }
}

let object = MyObject()
object.fullName
object.getName()
```

Even such a small example does quite a lot of work. Applying the `dynamic` keyword causes many issues:

- Dynamic message sending with `_objc_msgSend`
- Type casting
- Because we use the Objective-C dynamic dispatch method, we need to convert our "Name" Swift string type into NSString; when we get a result from that function call back into Swift code we need to do one more conversion from NSString back to Swift String
- No optimization and function call inlining
- Because methods are always dispatched dynamically, the Swift compiler can't do inline optimization or eliminate empty methods

A summary of avoiding Objective-C

You should avoid using Objective-C and its runtime behavior for achieving high performance.

You should avoid using the `dynamic` keyword at any price. You should almost never use it.

Inherit from a Objective-C class only when you need that class behavior, as for UIView. Use the `@objc` attribute only when you need to expose your type to Objective-C.

Summary

In this chapter, we have covered many topics related to Swift performance. First we need to understand what we need to improve and enable optimization settings to get the best performance.

Memory usage is very important for achieving high performance. First, we covered how using constants can have a positive impact on performance. The second and more important example reflects how using value types and structures reduces memory usage and also improves performance by using fast stack memory.

The third important topic we have covered is dispatching methods. We have analyzed and compared both Objective-C dynamic dispatch and Swift static dispatch. By looking at the assembly code we have seen how Swift actually performs method dispatch and how it can benefit performance.

We have also covered some dangerous operations that could decrease performance and that should be avoided.

In the next chapter, we will learn more about different data structures: their differences and performance characteristics.

5
Choosing the Correct Data Structure

In the previous chapter, we covered Swift-specific features that make Swift fast. It is no less important to choose the correct data structure for a specific use case. In this chapter, we will talk about different data structures, their differences, and when to choose one instead of the other.

In this chapter, we will cover these topics:

- An overview of data structures
- The Swift standard library collections
- Array, set, and dictionary
- Speeding up with Accelerate and Surge

An overview of data structures

Every programming language has built-in primitive data types, such as integer, double, character, string, and Boolean. The Swift programing language has some more complex types, such as enumeration, optionals, and tuples. By composing primitive types, we can build more complex data types. To compose them, we use structures and classes.

A data structure is a way of organizing data in a specific way so that it can be used efficiently for a specific task, for example, searching, checking for existence, and a quick update of values.

Collection types

Creating a new type and choosing the correct type for it, either a value or a reference, is an important task, which we covered previously in *Chapter 2, Making a Good Application Architecture in Swift*. There is a bigger impact on performance when we work with many instances of the same type when we put them into a collection. Choosing the correct collection for a particular task is very important.

Swift has some powerful building collections; we should take a look at them first.

Swift standard library collections

You will very often be using different collections to store and process data in your applications. Swift has three different built-in collection types: arrays, dictionaries, and sets.

The Swift standard library also has many functions for working with these collections, such as sort, find, filter, map, and many others. These functions have very efficient implementations, and you should use them instead of making your own. First, let's take a look at the different collections.

Arrays

An array is an ordered collection of values that provides access to its elements by indexes. It is a very simple and well-known collection. You would use an array in these situations:

- Simple element storage (often add/remove from the end)
- Elements need to be ordered
- Random access to elements

Arrays are usually implemented as a continuous block of memory in which you store values. Because memory blocks are usually located next to each other, access to elements can usually be transformed into simple pointer arithmetic: *third element = array start position + (2 * element size)*.

0	1	2	3	4
10	2	14	22	2

Using arrays

Arrays are perfectly fit for storing data for UITableView. Items need to be ordered. We need to know the number of items, get an item by its index, and be able to edit a collection. You would use an array when you need to store two or more objects of the same type.

Arrays are a very simple and flexible collection. But because of this, they are often overused in situations where we should use a set, or maybe a dictionary or some other custom collection.

Fast operations

Arrays have great performance for some operations with a constant complexity O(1), which doesn't increase with the size of the array. You can use them freely:

- **Accessing elements**: To access the elements, use these operations, `array[i]`, `array.first`, and `array.last`. The approximate time is 81 nanoseconds, or 0.000000081 seconds.

- **Appending an element**: To append an element, use this operation `array.append(i)`. The approximate time is 100 nanoseconds or 0,0000001 seconds.

Inserting and removing elements at both the beginning and a random place is also a very fast operation, but it has O(n) complexity. It increases with the size of the array, as shown here:

Array size	Time in seconds
0 to 50_000	0.00001
500_000	0.00019
5_000_000	0.0043
50_000_000	0.040
500_000_000	0.32

Slower operations

Some other operations on arrays increase very fast with increasing size of arrays. You should be careful while using such methods.

Search

Finding elements has O(n) complexity. The more the elements in the array, the more the time it will take to find out whether an element exists. To find out whether an element is present in the array, it has to iterate over every element and compare them:

```
let array: [Int]
let index = array.indexOf(3445)
```

If the search is not a primary operation that you perform on the collection and the size of the array is, for example, 500_000_000 elements, the search would take 0.5 seconds. If you have to perform searches very often and it's critical to do so very fast, use a set for search operations, or sort an array and use a more effective search algorithm, such as binary search.

Sort

Sorting has even bigger complexity than searching for an element; it has O(n * n) complexity. Sorting needs to iterate over an array to find the correct place for one element, and repeat it for every element. The `sort` standard library function has a very efficient implementation, and you should use it. It uses different sorting algorithms depending on the array size. Because sorting is expensive, you should cache the sorted result and reuse it if needed. Sorting an array of 500_000_000 `int` elements takes about 67 seconds.

Sets

A set is an unordered collection that stores unique objects. Sets are used for checking for the membership.

Usually, a set is implemented as a hash table. Elements in a set have to conform to **hashable** and **equatable** protocols.

When you add an element to a set or search for an element, it uses a hash function of an element to find an index for that element in the storage. Because of this, many operations on a Set are very fast, and it has O(1) complexity.

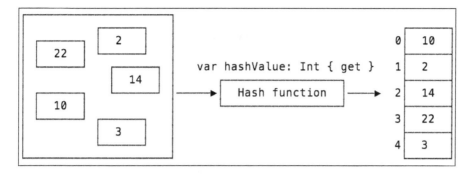

Using sets

Sets are great for checking whether an object exists in the collection. Also, they are great for getting the difference between two collections, for example, finding what objects were added or removed.

Sets have two important limitations. They are not ordered and can't contain duplicates:

```
var numbers: Set = [1, 1, 2, 3, 3, 4]
// {2, 3, 1, 4}
```

But, a set turns these limitations into features. Because of this and because it uses a hash table to store its elements, it achieves incredible performance for some operations with a constant complexity of O(1) that doesn't increase with the size of the set. The following are the operations:

- Insert:

```
numbers.insert(10)
```

- Lookup: `contains`, `IndexOf`, and `subscript`:

```
let number = numbers.contains(10)
let foundIndex =  numbers.indexOf(101)
let start: SetIndex = numbers.startIndex
let first = numbers[start]
```

- Delete: `remove` and `removeAtIndex`:

```
let number = numbers.remove(27)
let number = numbers.removeAtIndex(numbers.startIndex)
```

All of these operations take less than 6 microseconds, which is 0.000006 sec, even for a set with 50_000_000 elements.

If you are going to do a heavy search for elements, use Set. You can have both an array to store data and a Set for search operations. This will use twice as much memory, but the search will be instant.

Set operations

You can perform fundamental set operations on two different sets, such as combining, extracting, and getting common values.

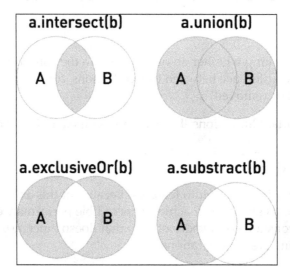

These operations are also very fast, but they have O(n) or O(log n) complexity, and with increased size, the time required to process the data increases:

```
let set = makeRandomSet(size)
let otherSet = makeRandomSet(set.count)

set.union(otherSet)
set.subtract(otherSet)
set.intersect(otherSet)
set.exclusiveOr(otherSet)
```

The performance of these operations is quite impressive, as you can see in the following table:

Set size	Union	Subtract	Intersect	ExclusiveOr
100 x 100	0.000015	0.000012	0.000013	0.00001
500_000 x 500_000	0.11	0.072	0.055	0.13

If we take a look at the declarations of these methods, we see that they accept, not a Set, but `SequenceType`:

```
func union<S : SequenceType where Element == Element>(sequence: S)
   -> Set<Element>
```

We can use these set methods with any `SequenceType`. Let's try to use an array instead of a set and see whether there will be any difference:

```
let set = makeRandomSet(size)
let otherSequence = makeRandomArray(set.count)

set.union(otherSequence)
. . .
```

Set size x array size	Union	Subtract	Intersect	ExclusiveOr
100 x 100	0.000013	0.0000045	0.000026	0.000032
500_000 x 500_000	0.10	0.058	0.13	0.18

As you can see, `intersect` and `exclusiveOr` perform better with a set. Although the difference is so small that it wouldn't have big impact on the overall application performance, it is still an important observation that you should remember.

Let's take a look at one more method in set— `isSubsetOf`:

```
func isSubsetOf<S : SequenceType where Element ==
Element>(sequence: S) -> Bool
```

It also has the `SequenceType` parameter, so it's possible to use both sets and arrays:

```
let set = makeRandomSet()

var otherSequence = Array(set)
otherSequence.append(random())
set.isSubsetOf(otherSequence)

var otherSet = set
```

```
otherSet.insert(random())
set.isSubsetOf(otherSequence)
```

The results are very interesting. With the set size equal to 5_000_000, `isSubsetOf` takes 4 minutes with an array argument, and less than 1 second with a set.

 For the `isSubsetOf` method, it is always preferable to use a set as an argument.

Size	isSubsetOf (array)	isSubsetOf (set)
50_000	0.11 sec	0.0045 sec
5_000_000	237.2 sec	0.46 sec

Dictionaries

A dictionary is an unordered collection that stores unique key-value pairs. Dictionaries are useful for quick object lookups by key

Dictionaries also use a hash table to store their keys and values. Because of this, a dictionary has similar performance characteristics as a set. Dictionaries are very useful when you need to connect two objects and perform instant searches and lookups for them.

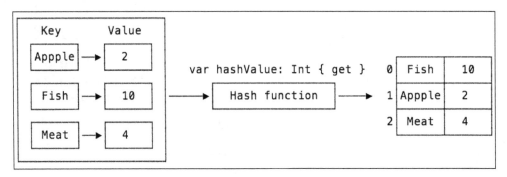

The dictionary collection is a very simple type. It doesn't have many methods of its own. The main functionality is to query a value for a key, update it, and delete it:

```
var capital = ["Germany" : "Berlin", "France"  : "Paris"]

capital["Norway"] = "Oslo"
capital.removeValueForKey("France")
capital["France"] = nil
```

```
if let index = capital.indexForKey("Spain") {
  print("found Spain at: \(index)")
}
capital.values
```

Using a dictionary would result in a big performance win in a situation where you have two arrays and you want to find a connection between them.

Collection memory allocation

Every collection has very similar performance optimization when you instance an instance of it. There are three different ways of creating an instance of a collection.

Empty

You can create an empty collection. All arrays, sets, and dictionaries have an empty `init` method:

```
let array = [Int]()
let set = Set<Int>()
let dic = [String : Int]()
```

Reserve capacity

The other way is to instance an instance of a collection and reserve a required memory capacity. All collections have dynamic size, and they allocate more memory when needed. When you know how many elements you are going to store in the collection, it is useful to allocate exactly the required amount of memory upfront:

```
var array = [Int]()
array.reserveCapacity(500_000)

var set = Set<Int>(minimumCapacity: 500_000)
var dic = [String : Int](minimumCapacity: 500_000)
```

Default values

An array has one more way of instantiating. You can create an array with default values set for all the elements in that array:

```
var array = [Int](count: 500_000, repeatedValue: 0)
array[i] = 10
```

Here are the results in seconds:

Size	Method	Array	Size set	Dictionary
500				
	Empty	5.2e-06	2.4e-05	2.4e-05
	Capacity	8.8e-07	1.6e-06	4.7e-06
	Default values	4.8e-07		
50_000_000				
	Empty	1.29	11.7	12.9
	Capacity	1.13	9.4	10.9
	Default values	1.043		

As you can see from the results, it is always better to reserve some capacity for collections when you know the size, even for small sizes.

For arrays, using default values is the fastest way, but you have to remember that the array is filled in with default values and you have to either handle them or replace them with real values.

Also, you can see that creating an array is an incredibly fast operation, even with big sizes. That's one of the advantages of arrays over other collections.

The CollectionType protocol methods

All the collections mentioned earlier — array, set and, dictionary — implement a `CollectionType` protocol. Because of this, they are interchangeable. You can use any of them in places where a `CollectionType` method is required. An example is a function with a `CollectionType` parameter:

```
func useCollection<T: CollectionType>(x: T) {
   print("collection has \(x.count) elements")
}

let array = [1, 2, 3]
let set: Set = [2, 2, 3, 4, 5]
let dic = ["A" : 1, "B" : 2]

useCollection(array)
useCollection(set)
useCollection(dic)
```

Protocol extensions

The other incredibly useful feature is protocol extensions. With protocol extensions, we can add implementations of methods and properties directly to the protocol. All types that conform to that protocol are able to use those methods for free. Let's add our own property to a `CollectionType` method:

```
extension CollectionType {
  var middle: Self.Index.Distance {
    return count / 2
  }
}

array.middle
set.middle
dic.middle
```

The types that conform to that protocol can provide their own implementation for that method. In such a case, this implementation will be used for that type instead of the one defined in the protocol extension:

```
extension Dictionary {
  var middle: Dictionary.Index.Distance {
    print("Dictionary middle")
    return count / 2 + 100 // :( wrong middle index
  }
}
dic.middle // 101 :)
```

The `CollectionType` protocol uses this functionality very heavily, and there are many methods and properties that are available for a `CollectionType`, for example, `isEmpty`, `first`, `dropFirst(n: Int)`, `map`, `indexOf`, and many more. Let's take a look at `isEmpty` and `dropFirst(n: Int)`:

```
array.isEmpty
set.isEmpty
dic.isEmpty

array.dropFirst(1)
set.dropFirst(1)
dic.dropFirst(1)
```

Let's inspect these methods. Hold down *command* and click on the `isEmpty` property to jump to its declaration. An array uses a `CollectionType` version of `isEmpty`, but set and dictionary provide their own implementations for the `isEmpty` property. Set and dictionary do this because the way in which they store their elements is different, and they can provide a better and more performance implementation of the `isEmpty` property. The types usually use their own implementation instead of using the protocol's default implementation if they can do it better.

You can read more about protocol extensions from the *The Swift Programming Language* book at `https://developer.apple.com/library/prerelease/ios/documentation/Swift/Conceptual/Swift_Programming_Language/Protocols.html`, in the *Protocols* section.

Accelerate and Surge

Both iOS and OS X SDK have a very powerful framework that provides high-performance functions for working with matrices, vectors, signals, image processing, and math operations. It is called the Accelerate framework. The Accelerate framework is quite big, so we will take a look at only one part that is related to working with collections; it is the vDSP part. You can find out more about it at `https://developer.apple.com/library/prerelease/ios/documentation/Accelerate/Reference/vDSPRef/index.html`.

First, let's implement the very basic mapping, calculating, and sum operations using the Swift standard library:

```
let array = [1.0, 2.0]
let result = array.map { $0 + 3.0 }
result // [4.0, 5.0]

let sum = array.reduce(0, combine: +)
sum // 3
```

This code is very clear and readable and doesn't need any comments. Let's try to do the same using Accelerate:

```
let array = [1.0, 2.0]
var result = [Double](count: array.count, repeatedValue: 0.0)

var add = 3.0
vDSP_vsaddD(array, 1, &add, &result, 1, vDSP_Length(array.count))
result // [4.0, 5.0]
```

```
var sum = 0.0
vDSP_sveD(array, 1, &sum, vDSP_Length(array.count))
sum // 3
```

As you can see, the code gets more complex than in the previous version. The vDSP library works with vectors and matrices. For the vDSP_vsaddD function call, we pass an input array. The second parameter gives the distance between the elements in the array. Because the elements in the array are next to each other, we use 1. The third is our second parameter, and the fourth and fifth are resulting arrays similar to the first and second parameters.

So, the Accelerate code is more complex, but it has better performance. Let's compare it:

```
let array = makeRandomDoubleArray(size)
var result = [Double]()

measure("map") {
  result = array.map { $0 + 3.0 }
}

// vDSP Version
let array = makeRandomDoubleArray(size)
var result = [Double](count : array.count, repeatedValue : 0.0)
var add = 3.0

measure("vDSP_vsaddD") {
  vDSP_vsaddD(array, 1, &add, &result, 1,
  vDSP_Length(array.count))
}
```

The results are very interesting for big arrays with 500_000_000 elements. The map function takes 5.1 seconds and vDSP_vsaddD takes 0.6 seconds. It runs almost 10 times faster.

Yes! Accelerate has way better performance, but the source code gets very complicated. However, there is a solution to it. We could make a nice API wrapper to work with the Accelerate framework. Fortunately, this has already been done. There is a very nice open source Swift framework called **Surge**. You can download it from GitHub at https://github.com/mattt/Surge.

After you have downloaded it, add the Surge framework project to your project. Add it as a linked framework, and then you can use it. Now, by using Surge, the code looks very nice and has great performance. Here is an example of calculating the sum of all elements using Surge. It's even nicer than using a reduce method:

```
import Surge
let numbers = makeRandomDoubleArray(size)
let sum = Surge.sum(numbers)
```

Other collections

We have covered three main Swift standard library collections. There are also other not-so-well-known and often-used helper collections, such as `ArraySlice`, `MapCollection`, `CollectionOfOne`, `ContiguousArray`, `EmptyCollection`, and `FilterCollection`. If you want to know more about them, just press *command* and click on any type. You will see the content of the Swift standard library. Then, just explore it!

You can also implement your own collections, if needed. The Swift generic allows you to make very abstract collections that can be used with any type. For example, it could be the Stack, or a Linked List, a Binary Tree or any other collection that fits your needs.

Summary

In this chapter, we covered the importance of choosing the correct data type. We covered the Swift standard library collections with their features and limitations. You learned which collection fits better for which use case. Also, we showed the performance characteristics when working with those collections and performing different operations. Moreover, we gave a tip on how to improve performance and memory allocation for collections.

In the next chapter, we will see how to create an application architecture that contributes to achieving high performance.

6
Architecting Applications for High Performance

In previous chapters, we talked about different ways of improving code to achieve high-performance. Mostly we concentrated on a small part of the code and how to improve a function, an algorithm, and a data structure. In this chapter, we will concentrate on higher levels. We will talk about how to design an application architecture that can be scalable, maintainable, and high-performance.

In this chapter we are going to cover these topics:

* High-performance and concurrency overview
* Avoiding state
* Divide and conquer
* Designing asynchronous architecture

Achieving high performance

One of the ways to improve application performance is to run code concurrently. This not only allows us to run code faster and get the results more quickly, but it also frees the main-thread from doing a lot of work and being blocked. You should know that the main thread is responsible for events and user input handling. All the UI work is performed on the main thread and to achieve a really smooth user interaction we should do as little work as possible on the main thread.

Running code concurrently can be a tricky task and sometimes it can lead to increased running time for an operation. Making solid concurrent application architecture is also not a trivial task and you should plan it carefully.

To take full advantage of concurrency, it is very useful to understand the hardware we have at our disposal that allows us to do that.

Device architecture

In order to be able to achieve really high-performance, first we need to learn and understand what kinds of tools we have at our disposal. If you are making an iOS application, your application will run on the iPhone and iPad; for OS X it would run on the Mac. Although it might seem that the iPhone and the Mac differ a lot, they share the same basic concept and we can think about a Mac as a more powerful iPad device.

Nowadays, all computers and even phones use multi-core processors that allow us to execute many instructions at the same time in parallel. Starting with the iPhone 4s, all iPhones have a dual-core processor and the iPad Air 2 even has a 3-core processor. We should fully use that power and take advantage of it.

Let's have a look at how we could design code that could be run in parallel on multi-core processors.

Concurrency overview

By default, when you make an application it runs the code in a single-thread environment, a main thread. For example, an iOS application would call the `application: didFinishLaunchingWithOptions` method on the main thread.

A simpler example is an OS X Command Line Tool application. It has only one file: `main.swift`. When you start it, the system creates a single main thread and runs all the code in the `main.swift` file on that thread.

For testing code, playgrounds are the best. By default, playgrounds stop after executing the last line of code and don't wait for the concurrent code to finish executing. We can change this behavior by telling the playgrounds to keep running indefinitely. To do that, include these two lines in the playground file:

```
import XCPlayground
XCPSetExecutionShouldContinueIndefinitely()
```

Now we can start playing with concurrency. The first task we need to do to run code concurrently is to schedule a task to be run on a different thread. We can schedule a task for concurrent execution by using:

- Threads
- **GCD (Grand Central Dispatch)**
- Operation Queues

Threads

As the first option, we could use threads. A thread is the most low-level API. All the concurrency is built on top of threads and runs multiple threads. We can use NSThread from the Foundation framework. The simplest way to do this is to create a new class with a method that will be the starting point for our new thread.

Let's see how we could schedule the new thread:

```
class Handler: NSObject {
  @objc class func run() {
    print("run")
  }
}

NSThread.detachNewThreadSelector("run", toTarget: Handler.self,
  withObject: nil)

let thread = NSThread(target: Handler.self, selector: "run",
  object: nil)
thread.start()
```

You can create a new thread in two ways, by using the detachNewThreadSelector function or create an instance of NSThread and use the start function. We have to mark our run function with the @objc attribute because we use it as a selector when creating a thread, and NSThread is an Objective-C class that uses dynamic dispatch for method calling.

The NSObject has a simple API for performing a method on a different thread. Because our handler inherits for NSObject we can use it.

```
Handler.performSelectorInBackground("run", withObject: nil)
```

Another way is to create a subclass of `NSThread` and override the starting point of a thread, the `main` function. In that way we don't need a handler class.

```
class MyWorker: NSThread {

  override func main() {
    print("Do work here")
  }
}

let worker = MyWorker()
worker.start()
```

Thread complexity

Even though the code is pretty simple here, working with threads is quite a complex operation. We need to take care of managing the state of the thread, correctly terminating it, and releasing the resources used by the thread.

Creating a new thread is a very expensive and time-consuming operation, and we should avoid it when possible. The way to solve this is to reuse created threads. Creating and managing a thread-pool adds extraordinary complexity to the application that we don't need.

The process gets even harder when you need to communicate between threads and synchronize data between them.

Solution for threads

Instead of solving our initial task that we wanted to run concurrently, now we are spending time managing the complexity of that concurrent execution system. Fortunately we don't need to do that as there is a solution: *Don't use threads*.

The *iOS and Mac Concurrency Programming Guide* recommends not using threads but choosing a high-level API, a GCD, or Operation Queues.

 Thread APIs are shown in this chapter only for general knowledge. You should almost never use threads; use GCD instead.

GCD

GCD (Grand Central Dispatch) is a high-level API that is built on top of threads and performs all aspects of thread management for you. Instead of working with threads, GCD provides a queue and task abstraction. You schedule a task to a queue for execution and the queue takes care of everything else. Let's see how we could rewrite our code with GCD:

```
let bgQueue = dispatch_get_global_queue(QOS_CLASS_BACKGROUND,0)
dispatch_async(bgQueue) {
  print("run")
}
```

As you can see, the code looks simpler from the start. Before we dive into the details, let's have a look at GCD and its concepts:

- Queues
- Tasks
- Adding tasks to the queue

Queues

A **queue** is a structure that takes care of managing and executing its tasks. The queue is a first-in first-out data structure. That means that tasks in the queue are started in the order they were added to the queue.

 First in first out means that tasks are started in the same order but it doesn't mean that they can't be executed simultaneously. Concurrent queues can start many tasks at the same time.

The queue itself doesn't have much functionality. The main operation you would need to do is to create a queue or get one of the global queues.

There are three queue types:

- Main queue
- Concurrent: global and own queues
- Serial

Main queues

A **main queue** represents a queue associated with a main thread. It runs tasks serially, one after the other. You would usually use this queue to pass the result of an execution from other background queues to the main queue to update the UI state. You can get a main queue by calling the `dispatch_get_main_queue` function.

```
let mainQueue = dispatch_get_main_queue()
```

Concurrent queues

A **concurrent queue** runs its tasks concurrently. The easiest way to get a concurrent queue is to use a global concurrent queue.

```
func dispatch_get_global_queue(identifier: Int, flags: UInt) ->
dispatch_queue_t!
```

To get a global queue, we need to specify what kind of priority we need. There are five types of queue with descending task priority. `USER_INTERACTIVE` is the most prioritized queue and `BACKGROUND` is the least.

- `QOS_CLASS_USER_INTERACTIVE`
- `QOS_CLASS_USER_INITIATED`
- `QOS_CLASS_DEFAULT`
- `QOS_CLASS_UTILITY`
- `QOS_CLASS_BACKGROUND`

 Also available are old `DISPATCH_QUEUE_PRIORITY` constants that can be used when specifying a queue priority type instead of `QOS_CLASS` but `QOS_CLASS` is preferred.

The second flag parameter is reserved and never used, so we just use 0. The global queues are available for use by the whole system and everyone can add tasks to them. When all you need is to run some tasks, this is a perfect fit.

Own queues

When you need to do more complex handling and have full control of what tasks are added to the queue, you can create your own queue. Own queues fit well when you need to be notified of when all tasks are done, or to do more complex synchronization between tasks.

You can create both concurrent and serial queues. Serial queues execute one task at a time, one after another, not concurrently.

```
let concurentQ = dispatch_queue_create("my-c",
DISPATCH_QUEUE_CONCURRENT)
let serialQ = dispatch_queue_create("my-s", DISPATCH_QUEUE_SERIAL)
```

Tasks

A **task** is a block of code that needs to be run. A task is defined as a `dispatch_block_t` type and it is defined as `() -> Void`. We could use a closure or a function as a task.

```
typealias dispatch_block_t = () -> Void

let tasks: dispatch_block_t = {
  print("do Work")
}

func doWork() {
  print("do Work Function")
}
```

Adding tasks to the queue

We have a queue and we have a task that we want to run. To run a task on a particular queue, we need to dispatch it to that queue. We could do this in two ways:

- **Synchronous**: `dispatch_sync`
- **Asynchronous**: `dispatch_async`

Both functions are very simple and have the same type. The only differences are in their names and the way they work.

```
dispatch_sync(queue: dispatch_queue_t, _ block: dispatch_block_t)
dispatch_async(queue: dispatch_queue_t, _ block: dispatch_block_t)
```

Synchronous dispatch

Synchronous dispatch submits a task for execution and waits until the task is done.

```
dispatch_sync(queue) { ... }
```

When you use a concurrent queue and dispatch a task to it synchronously, the queue can run many tasks at the same time, but the `dispatch_sync` method waits until the task you submitted is finished.

```
let queue = dispatch_get_global_queue(QOS_CLASS_BACKGROUND, 0)

dispatch_sync(queue) { print("Task 1") }
print("1 Done")

dispatch_sync(queue) { print("Task 2") }
print("2 Done")
```

> Never call the `dispatch_sync` function from a task that is executing in the same queue. This would cause a deadlock for the serial queue and should be avoided for concurrent queues as well.
>
> ```
> dispatch_sync(queue) {
>   dispatch_sync(queue) {
>     print("Never called") // Don't do this
>   }
> }
> ```

In this example, the `print("1 Done")` instruction and the rest of the code will wait until `Task 1` is finished.

Asynchronous dispatch

Asynchronous dispatch, on the other hand, submits a task for execution and returns it immediately.

```
dispatch_async(queue) { ... }
```

If we use the previous example and change it to use `dispatch_async`, `1 Done` will not wait for tasks to be finished. We can also simulate extra work by freezing the current thread with a sleep function.

```
let queue = dispatch_get_global_queue(QOS_CLASS_BACKGROUND, 0)

dispatch_async(queue) {
  sleep(2) // sleep for 2 seconds
  print("Task 1")
}
print("1 Submitted")

dispatch_async(queue) { print("Task 2") }
print("2 Submitted")
```

As a result, `Task 2` is submitted for execution immediately after `Task 1` and it finishes execution before `Task 1`. Here is the console output:

`1 Submitted`

`2 Submitted`

`Task 2`

`Task 1`

GCD also has some powerful tools for synchronizing submitted tasks, but we are not going to cover them here. If you want to learn more, read the *Concurrency Programming Guide* article in the Apple library documentation at `https:// developer.apple.com/library/ios/documentation/General/Conceptual/ ConcurrencyProgrammingGuide`.

Operation queues

`NSOperationQueue` is built on top of GCD and provides more high-level abstraction and an API that allows us to control an application controlflow.

The concept is very similar to GCD; it has a queue and tasks that you add to the particular queue.

```
let queue = NSOperationQueue()
queue.addOperationWithBlock {
  print("Some task")
}

NSOperationQueue.mainQueue().addOperationWithBlock {
  print("Some task")
}
```

The NSOperationQueue provides a more high-level API but it is also a bit slower than GCD. NSOperationQueue fits very well with controlling the application flow, when tasks need to be chained, depend on each other, or need to be canceled. You can achieve the same functionality by using GCD but it would require implementing some extra logic that is already implemented by the NSOperationQueue.

GCD works very well when you need to perform a task and get the result and do not need to control the application flow.

Further in this chapter we will use GCD for concurrency. Now, let's move on and learn some tricks that will help us to make our code architecture more solid for concurrent programming.

Designing asynchronous code

The first characteristic of asynchronous code is that it returns immediately and notifies the caller when it has completed the operation. The best solution is to return the result of the computation as well. This way we get more function style *Input -> Output* functions style.

Let's have a look at this simple example. This code has many issues and will refactor them all.

```
class SalesData {

    var revenue: [Int]
    var average: Int?

    init (revenue: [Int]) {
      self.revenue = revenue
    }

    func calculateAverage() {

      let queue = GCD.backgroundQueue()
      dispatch_async(queue) {

        var sum = 0
        for index in self.revenue.indices {
        sum += self.revenue[index]
        }

        self.average = sum / self.revenue.count
      }
    }
}
```

We have made a GCD structure that provides a nice API for working with GCD code. In the preceding example we have used a GCD backgroundQueue function; here is its implementation:

```
struct GCD {
    static func backgroundQueue() -> dispatch_queue_t {
      return dispatch_get_global_queue
      (QOS_CLASS_BACKGROUND, 0)
    }
}
```

On the whole, the calculation code in that example is really bad and we could improve it by using a `reduce` method that would actually solve many problems and make the code safer and more readable.

```
let sum = self.revenue.reduce(0, combine: +)
```

But the main point of that example was to show how dangerous it could be and what kinds of issues you could face with this architecture.

Let's use this code to see the problem:

```
let data = SalesData(revenue: makeRandomArray(100))
data.calculateAverage()
data.average // nil
```

The problem is that `calculateAverage` returns immediately as it is supposed to and the average is not calculated at this moment. To solve that problem all the asynchronous code should have some way to notify a caller when the task is completed. The easiest way to do this is to add a callback completion function as a parameter.

```
func calculateAverage(result: () -> Void ) {
    . . .

    self.average = sum / self.revenue.count
    result()
}
```

Now, when using this code, we can use a nice and clear trailing closure syntax for the result callback parameter.

```
let data = SalesData(revenue: makeRandomArray(100))
data.calculateAverage {
    print(data.average)
}
```

There is one very important issue with this code. It is calling the `result` callback function on the background thread. It means that the closure we pass to `data.calculateAverage` will be called on the background but for us it's not documented and this behavior is not clear. We suppose that we will get that closure called on the main thread, because we are calling the `calculateAverage` function from the main thread. Let's do that. What we need to do is to switch to the main queue and call `result` on the main thread.

```
dispatch_async(GCD.mainQueue()) {
    result()
}
```

The best practice is to always call a callback method on the main queue by default if another behavior is not required. If you need to call a callback on the special queue, then it should be passed to a function as a parameter.

This code works but there is still one improvement that could be done. When the result callback gets called, the first thing we do is get the average instance. It would be way better if the result callback returned the result of its computation.

 In general terms, it is a good functional programming style for functions to take input and return the result, X -> Y. These functions are easier to use and test.

Let's refactor this code to pass an average result number to the callback function:

```
func calculateAverage(result: (Int?) -> Void ) {
  ...

  self.average = sum / self.revenue.count

  dispatch_async(GCD.mainQueue()) {
    result(self.average)
  }
}

// Use case
let data = SalesData(revenue: makeRandomArray(1000))
data.calculateAverage { average in
  print(average)
}
```

The change is not big but the benefits are quite extensive. When we use the calculateAverage function we get the result directly in the closure as a parameter. Now we don't need to access the instance variable of SalesData. SalesData becomes more of a closed-box type with hidden implementation details and because of that we will be able to do more refactoring in the future.

Avoiding state

The first rule is to avoid a state. When you are doing two things at the same time, those two processes should be as independent and isolated as possible. They shouldn't know anything about each other or share any mutable resources. If they do, then we would need to take care of synchronizing access to that shared resource, which would bring a complexity to our system that we don't want. Right now in our code we have two states: a `revenue` numbers array and the `average` result. Both of the processes have access to that state.

The first problem in that the code is referencing itself. When you try to access an instance variable or a method that is out of the closure scope, you see an error message: **Reference to property 'revenue' in closure requires explicit 'self.' to make capture semantics explicit**.

Xcode would also propose a fix to this issue, adding explicit self-capturing. This would solve the Xcode error but it wouldn't solve the root problem. When you see this error, stop and rethink your code design; in some cases it would be better to change the code, like in our case.

```
26    dispatch_async(queue) {
27        let sum = self.revenue.reduce(0, combine: +)
28        self.average = sum / self.revenue.count        Referen
```
Issue ⊙ Reference to property 'revenue' in closure requires explicit 'self.' to make capture semantics explicit

Fix-it Insert "self."

The second problem is having a mutable state and mutating an instance variable. Let's use our last example once more and see why it's a bad idea to have a state and access instance variables in the concurrent code:

```
let data = SalesData(revenue: makeRandomArray(1000))
data.calculateAverage { average in
  print(average)
}
data.revenue.removeAll(keepCapacity: false)
```

If you run this code, it will crash with a **fatal error: Array index out of range** error due to getting the number from the array by an index operation:

```
sum += self.revenue[index]
```

What is happening here is that, when we call `calculateAverage`, the revenue array has data, but later we remove all the revenue numbers and the arrays become empty; however, the indexes we are iterating over point to an old array size, and we are trying to access the index out of the bound arrays.

To solve that problem we should always try removing a state. One way to do that is to pass the needed data to a function as arguments or, if you want to have some state as in our case, capture the immutable values for a closure.

Capture list

The first step to make this code better is to remove accessing mutable array variables. The easiest way to solve this is to make a local constant and use it in the closure.

```
let revenue = self.revenue
dispatch_async(queue) {
  var sum = 0
  for index in revenue.indices {
    sum += revenue [index]
  }

  self.average = sum / revenue.count
  ...
}
```

This solution would work, because modifying the `revenue` instance variable doesn't have an impact on the local constant we have created. This code has one small issue. The constant is visible outside the closure, but it's intended to be used only inside the closure. It would be better if we could move it to the closure. We can do this by using a capture list of a closure. The only one change we need to do is to remove the local constant declaration and add it to the closure capture list. The rest of the code stays the same.

```
dispatch_async(queue) { [revenue] in
  ...
}
```

In this example, we used a very short capture list notation, but we could also provide an alternative name for the constant we are capturing and add additional ARC weak or unowned attributes.

```
dispatch_async(queue) { [numbers = revenue] in  ...
```

Immutable state

Having a state in the concurrent code is a bad design but there are two types of state: mutable and immutable. In any case, you will need to have some sort of a state in the application. If you are going to have a state, make it immutable; in that way you will ensure that it won't be changed and you can safely work with it.

If we have a look at our previous code example we could make the `revenue` numbers immutable, which would solve the problem:

```
class SalesData {

    let revenue: [Int]
    ...
}
```

The first small change we would do is to change the revenue number array to be immutable. Because the `revenue` array is immutable it's not possible to modify it after we created an instance, so we need to remove this code.

```
data.revenue.removeAll(keepCapacity: false)
```

Because `revenue` is immutable now, it's safe to use it in a concurrent code, so we can remove the capture list and use the immutable revenue directly by using `self` explicitly as we did before.

```
func calculateAverage(result: (Int?) -> Void ) {

    let queue = GCD.backgroundQueue()
    dispatch_async(queue) {

        var sum = 0
        for index in self.revenue.indices {
            sum += self.revenue[index]
        }

        self.average = sum / self.revenue.count
        dispatch_async(GCD.mainQueue()) {
            result(self.average)
        }
    }
}
```

`SalesData` contains immutable sales numbers that cannot be changed. This means that, after we have calculated the average value once, it will be the same all the time for that instance. The next time we call `calculateAverage`, we don't need to calculate it again if we can reuse it.

```
func calculateAverage(result: (Int?) -> Void ) {

    if let average = self.average {
        result(average)
    } else {
```

```
        let queue = GCD.backgroundQueue()
        dispatch_async(queue) {
          ...
        }
      }
   }
```

We can even carry out one more step to make it immutable and use `struct` instead of a `class` for the `SalesData` type. When we do this, we will get an error saying:

cannot assign to property: 'self' is immutable

self.average = sum / self.revenue.count

When you assign new values to `self.average`, you are modifying a self instance, and because structs are immutable by default we need to mark that method as mutating:

```
   struct SalesData {
     ...
     mutating func calculateAverage(result: (Int) -> Void ) {
       ...
     }
   }
```

Those are only two changes we need to do. Also, when we are using it, we need to make an instance of `SalesData` as a variable, because `calculateAverage` is mutating it.

```
   var data = SalesData(revenue: makeRandomArray(1000))
   data.calculateAverage { average in
     print(average)
   }
```

So now we can't have a constant `let` immutable `SalesData` instance. This is not a sign of good architecture and we should refactor it. Using a struct for data entities is a very good solution so we should keep refactoring code with this approach.

Divide and conquer

The best way to achieve good, solid application architecture is to structure code well, create appropriate abstractions, and separate it into components with a single responsibility.

In functional programming it goes even further. The data — and the functions to work on that data — are also separated. The OOP concept of data and methods to work with it are split into two parts. This makes code even more flexible and reusable.

For concurrent code execution it's particularly important to split your code into standalone separate pieces because they can be sent for execution concurrently without blocking each other.

Before we start refactoring the code let's analyze it first. The goal is to identify a component with a single responsibility. I did it and here are those components:

- **In Data**: The first part is our input data. In our case it is a `SalesData` structure that holds immutable data in our application.

- **Calculation function**: The next part is our function that knows how to calculate the average for `SalesData`. It's a simple first-class function that takes `SalesData` and returns the average. Its mathematical notation would be `f(x) = y` and the code notation would be
`func calculateAverage(data: SalesData) -> Int`.

- **Result data:** This is a result returned by the calculation function. In our example, it is a simple `Int` number that represents an average.

- **Async execution operation**: The next part is an operation that switches to the background thread and back to the main thread and that actually allows us to perform asynchronous code execution. In our example it's a `dispatch_async` function call with an appropriate queue.

- **Cache**: Once we have calculated an average, we store it and don't perform the calculation again. This is exactly what a cache is for.

Now we have identified separate components in our application, let's build relations and communication between them.

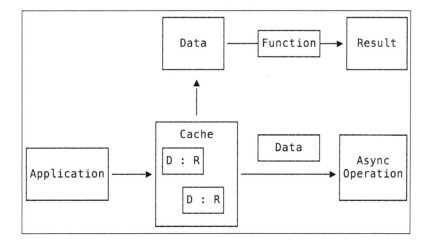

To keep interaction simple, our application will ask a cache for an average value of
`SalesData`. If a cache contains an average value, it will return it. Otherwise, it will
start an async operation and pass `SalesData` to it. The async operation will call a
`calculateAverage` function, get an average result, and pass it back to the cache.
The cache will save it and forward it to the application.

This might sound a bit complicated when it's explained in words, but in code it's
pretty simple, straightforward, and clear. Before we begin refactoring, let's have a
look at the code that we made this structure for:

```swift
struct SalesData {
   let revenue: [Int]
   var average: Int?

   init (revenue: [Int]) {
     self.revenue = revenue
   }

   mutating func calculateAverage(result: (Int?) -> Void ) {

     if let average = self.average {
       result(average)
     } else {

       let queue = GCD.backgroundQueue()
       dispatch_async(queue) {
         var sum = 0
         for index in self.revenue.indices {
           sum += self.revenue[index]
         }

         self.average = sum / self.revenue.count
         dispatch_async(GCD.mainQueue()) {
           result(self.average)
         }
       }
     }
   }
}
```

The first idea that came to my mind was to follow the FP principle *keep data and functions separate* and move the `calculateAverage` function outside a `SalesData` type.

```swift
struct SalesData {
   let revenue: [Int]
   var average: Int?
}

func calculateAverage(data: SalesData, result: (Int) -> Void ) {
   ...
}
```

This would work, but there is one issue with this code. The `calculateAverage` function can only work with the `SalesData` type, so it should be hidden inside the `SalesData` type and not be visible to other types. Also, in the Swift method notation is preferred.

Swift 2.0 moves to methods over free functions, so it prefers to use an immutable method.

Example:

- **Swift 2.0 Methods**: `[1,2,3].map { $0 + 1 }.filter { $0 > 1 }`
- **Swift 1.2 Function**: `filter(map([1,2,3]) { $0 + 1 }) { $0 > 2 }`

Instead of moving the `calculateAverage` function out of the `SalesData` type, let's make it immutable and make it only perform an average calculation instead, as we have shown in our schema.

`SalesData` should:

- Store revenue numbers
- Be an immutable function for calculating its average

Let's refactor the `SalesData` structure and remove all other methods to follow our new structure

```swift
struct SalesData {
   let revenue: [Int]

   var average: Int {
      return revenue.reduce(0, combine: +) / revenue.count
   }
}
```

Here is the solution and it's very clean and simple. Instead of a function we have used a computed property. Swift tends to use more read-only properties for immutable data and in our example it will make for better readability in the future. Also, we have finally used the reduce method for calculating the average. We can use it in this way:

```
let data = SalesData(revenue: [145, 24, 3012])
data.average
```

The next step is to make it work asynchronously. Let's make a new type for it. It should take a SalesData and a callback closure, which will return Int, a calculated average result.

```
struct AsyncOperation {

  static func calculate(data: SalesData, result: (Int) -> Void ) {
    GCD.asyncOnBackground {
      let average = data.average
      GCD.asyncOnMain {
        result(average)
      }
    }
  }
}
```

We have added two more helper methods to our GCD type:

```
struct GCD {
  static func asyncOnBackground(block: () -> Void ) {
    dispatch_async(self.backgroundQueue(), block)
  }
  static func asyncOnMain(block: () -> Void ) {
    dispatch_async(self.mainQueue(), block)
  }
}
```

This code looks okay, but there is one more issue with it. Calling an average is embedded together with switching to the background and main threads. It would be better if we kept these functions separate, so they could be reused if we wanted to add growth numbers and do a similar calculation for them.

```
static func calculateAverage(data: SalesData, result: (Int) ->
Void ) {
  runAsync(data.average, result: result)
}
```

```
//MARK: - Private
private static func runAsync<T>(@autoclosure(escaping) work: () ->
  T, result: (T) -> Void ) {
  GCD.asyncOnBackground {
    let x = work()
    GCD.asyncOnMain {
      result(x)
    }
  }
}
```

Here we created a `runAsync` generic function that performs some work on the background, and returns its result on the main thread. We have used an `@autoclosure(escaping)` attribute here in order to be able to pass an expression `data.average, ...`) instead of wrapping it into a closure manually. This makes the code syntax cleaner.

Now we can calculate the average in an asynchronous way.

```
let data = SalesData(revenue: [145, 24, 3012])
AsyncOperation.calculateAverage(data) { average in
  print(average)
}
```

Now it's time to build our last component, a cache. For the caching functionality a dictionary would be the best choice. Let's add a dictionary to store average results for `SalesData`.

```
struct SalesDataCache {
  var revenueCache = [SalesData : Int]()

  subscript (data: SalesData) -> Int? {
    return revenueCache[data]
  }

  mutating func getAverage(data: SalesData, result: (Int) -> Void)
  {
    if let average = self[data] {
      result(average)
    } else {
      AsyncOperation.calculateAverage(data) { average in
        self.revenueCache[data] = average
        result(average)
      }
    }
  }
}
```

We have created a `SalesDataCache` struct with one property, a cache, and a function to get the average value that either gives a cached value or calculates it and then saves it to the cache and returns. A very simple solution, but it won't work. It shows an error: **Type 'SalesData' does not conform to protocol 'Hashable'**.

The keys in the dictionary have to be `Hashable`, so we need to implement this in our `SalesData` type. The `Hashable` protocol requires that we implement the `hashValue` property and the equality function.

```
var hashValue: Int { get }
func ==(lhs: Self, rhs: Self) -> Bool
```

Implementing a good hash function for an array of numbers is quite complex. The easiest way to do it is to add an `id` property to `SalesData` that will uniquely identify it.

```
struct SalesData {
   let id: Int
   revenue: [Int]
}

//MARK:- Hashable
extension SalesData: Hashable {
   var hashValue: Int {
     return id.hashValue
   }
}

func == (lhs: SalesData, rhs: SalesData) -> Bool {
   return lhs.id == rhs.id
}
```

Now our cache will work and we can finally use it in our application. Let's do that:

```
let range = 0...10
var cache = SalesDataCache()
let salesData = range.map {
   SalesData(id: $0, revenue: makeRandomArray(1000))
}

for data in salesData {
   cache.getAverage(data) { average in
     print(average)
   }
}
```

As you can see, the API we have created is really easy to use. Even though there is a lot of logic going on behind the scenes, for you it's as simple as calling one method: getAverage.

Also, we have structured the underlying components in such a way that they can be used separately — for example, if we don't need a cache or asynchronous execution.

To wrap up the refactoring work on this example, let's see the full code we have ended up with:

```swift
struct SalesData {
   let id: Int
   let revenue: [Int]

   var average: Int {
      return revenue.reduce(0, combine: +) / revenue.count
   }
}

//MARK:- Hashable
extension SalesData: Hashable {
   var hashValue: Int {
      return id.hashValue
   }
}

func == (lhs: SalesData, rhs: SalesData) -> Bool {
   return lhs.id == rhs.id
}

struct AsyncOperation {

   static func calculateAverage(data: SalesData, result: (Int) ->
   Void) {
     runAsync(data.average, result: result)
   }

   //MARK: - Private
   private static func runAsync<T>(@autoclosure(escaping) work: ()
   -> T, result: (T) -> Void ) {
     GCD.asyncOnBackground {
       let x = work()
       GCD.asyncOnMain {
         result(x)
       }
     }
```

```
        }
      }
    }

  struct SalesDataCache {
    private var revenueCache = [SalesData : Int]()

    subscript (data: SalesData) -> Int? {
      return revenueCache[data]
    }

    mutating func getAverage(data: SalesData, result: (Int) -> Void)
    {
      if let average = self[data] {
        result(average)
      } else {
        AsyncOperation.calculateAverage(data) { average in
          self.revenueCache[data] = average
          result(average)
        }
      }
    }
  }
}
```

Controlling the lifetime

In our code, we have used an @autoclosure(escaping) attribute. It is a very powerful attribute and it deserves to be covered in detail. There is also an @noescape attribute. Let's explore them in more detail.

Applying the @autoclosure and @noescape attributes

First, let's have a look at when and how we could use these attributes. We can apply them to a function parameter with a function type. A function type can be represented as a method, function, or closure and it has (parameters) -> (return) notation. Here are a few examples:

```
func aFunc(f: () -> Void )
func increase(f: () -> Int ) -> Int
func multiply(f: (Int, Int) -> Int ) -> Int
```

@autoclosure

The `@autoclosure` attribute can be applied to a parameter with a function type that has no arguments and returns any type, `()` `->` `T`. For example:

```
func check(@autoclosure condition: () -> Bool)
func increase(@autoclosure f: () -> Int ) -> Int
```

When we use an `increase` function without the `@autoclosure` attribute, we need to pass a function, a method, or a closure as a parameter.

```
let x = 10
let res = increase( { x } )
check( { x > 10 } )
```

But in this use case it would be better if we could simply use an expression without the need to wrap it in a closure, like this:

```
let x = 10
let res = increase(x)
check(x > 10)
```

And that's exactly what `@autoclosure` allows us to do. When you make a parameter with the `@autoclosure` attribute, the expression you pass as an argument is automatically wrapped into a closure for you. It makes your code cleaner. That's all it does. No magic; it simply removes boilerplate code for you.

@noescape

The `@noescape` keyword is more complex and interesting. It can be applied to a function parameter with any function type.

The `@noescape` attribute indicates that a closure will be used inside a function body, before the function return is called. It means it won't escape the function body.

When you apply this attribute, it indicates that a closure will be used synchronously inside the function body. Also, it means that it will be released when you leave the function. The lifetime of that closure parameter can't outlive the function call.

Applying this attribute enables some performance optimization but, more importantly, it disables the requirement to explicitly specify `"self."` when accessing instance members.

Let's have a look at some examples to better understand this. For a simple example, we will use the same `increase` function, but now we will make it a method of a `struct`:

```
func increase(f: () -> Int ) -> Int {
   return f() + 1
}

struct Data {
   var number: Int

   mutating func increaseNumber() {
      number = increase { number }
   }
}
```

The `increase` function call contains an error: **Reference to property 'number' in closure requires explicit 'self.' to make capture semantics explicit**; we need to explicitly reference `self.number`.

But let's have a look at the `increase` function. The `f: ()-> Int` parameter is used inside the function body and it's not leaving its scope. This is a great candidate for applying the `@noescape` attribute to it.

```
func increase(@noescape f: () -> Int ) -> Int {
   return f() + 1
}
```

Now we don't need to do any further changes and explicitly reference `self.numbers`, because `@noescape` guarantees that a closure will be called before we leave that function and we can safely reference `self`.

 Apply `@noescape` wherever possible. It adds an extra security level to the code. Also, it enables better optimization and increases performance.

If we have a look at methods and function such as `map`, `reduce`, `contains`, and others in the Swift standard library, you will see that they are marked with the `@noescape` attribute. The golden rule is: *If you call the closure parameter before you leave the function, mark it with @noescape.*

Maybe in the future Swift will automatically do this for you, but for now we need to do it ourselves.

@autoclosure (escaping)

The @autoclosure attribute also applies an @noescape implicitly. If you want to make a parameter an autoclosure, while indicating that it will have a bigger lifetime than a function, use an @autoclosure(escaping) attribute. It could be useful for asynchronous code execution, like in our example with AsyncOperation.

Summary

In the first part of this chapter, we covered multithreading concurrency and multi-core device architecture. This general information allows us to understand the core principles of concurrent code execution.

In the second part, we covered three ways to run code asynchronously in Swift, by using threads, GCD, and NSOperation. We have explored the differences between them and the situations for which each is most suitable.

In the third part of the chapter, we concentrated on architecting asynchronous Swift code by using GCD. We have covered important tips such as passing a callback function parameter, avoiding a state, using immutable values, and others. Also, we have covered two Swift attributes — @noescape and @autoclosure — that are very useful.

In the next chapter, we will cover one more important performance optimization technique: Lazy Loading.

7
The Importance of Being Lazy

Another way of improving an application's performance is by deferring the execution of code until the result is needed. This sounds very logical; the less the code we run, the less the time it will take. This design pattern is usually called **lazy**. There are many things that can be lazy, and there are many different ways in which we can defer code execution. We are going to cover the following topics in this chapter:

- The lazy mindset
- Lazy loading
- Lazy collections and evaluation

The lazy mindset

First, it is very important to understand the lazy pattern, how it works, how it could benefit an application's performance, and when to use it. It's also very important not to abuse it, because that would make code more complex and hard to read. Plus, it would be hard to follow the execution flow. Also, abusing it would decrease the overall application's performance.

The general idea of a lazy pattern is to defer the evaluation of an instruction until someone asks for the result of that instruction.

In general, code is executed instruction by instruction, starting at the top of a file or a function. Nowadays, our applications are more complex and consist of many files, windows, libraries, components, and layers, but they still execute code in the same way. Because our system gets bigger and bigger, it's important to make them lazy so that we don't have to perform all the work when we start the application. Let's learn a few techniques for making code lazy.

Separation

It's very important to separate code into components with a **single responsibility** pattern. A component should do only one thing and do it well. Let's take a look at this simple example to understand how this could improve performance:

```
struct Person {
    let name: String
    let age: Int
}

func analyze(people: [Person]) {
    let names = people.map { $0.name }
    let last = names.maxElement()

    let alphabetOrder = names.sort { $0 > $1 }
    let lengthOrder = names.sort { $0.characters.count <
        $1.characters.count }
    let longestName = lengthOrder.last

    print(last, alphabetOrder, lengthOrder, longestName)

    let age = people.map { $0.age }
    let youngest = age.minElement()
    let oldest = age.maxElement()
    let average = age.reduce(0, combine: +) / age.count

    print(youngest, oldest, average)
}

let people = [Person(name: "Sam", age: 3),
    Person(name: "Lisa", age: 68),
    Person(name: "Jesse", age: 35)
]

people + EnglandPopulation()
analyze(people)
```

The problem with this code that the `analyze` function does two things. Because of that, the memory usage peak is twice as big. When we are done analyzing names, the memory used for it is not released immediately, but is kept until we return from the function. If we tried to analyze all of the population of England, it would require quite a lot of memory. By splitting them into separate functions, we can improve memory usage:

```
func analyze(people: [Person]) {

  let names = people.map { $0.name }
  analyzeNames(names)

  let age = people.map { $0.age }
  analyzeAge(age)
}

func analyzeNames(names: [String]) {
  ...
}

func analyzeAge(age: [Int]) {
  ...
}
```

You should apply a code separation technique very often to different components as an example. Make simple models and use only the data that you need for the current view. Splitting code into smaller components is useful, but you shouldn't split it into too small components.

Do work on demand

The main idea of lazy code is to do work only on demand. In this way, you can delay a code execution or remove it completely if no one asks for the result. When you are planning an application's architecture, try to think about these questions: when would this resource or data be used in the application? Does it have to be instantiated now or can it wait? How often would this data be used? If often, then maybe you should cache it; if not, maybe you should load it lazily. Is the data heavy? Do you need to clean it afterward? Asking all of these questions helps you build a better application architecture with the *do work on demand* approach.

Approximate results

Another idea is mixing lazy loading and asynchronous execution. If you want to perform an asynchronous task, because it takes a long time, but you would also like to get a result immediately, you can return an approximate result immediately and continue the execution. For example, the core data asynchronous fetching implements this pattern. It returns you an approximate number of items that will be fetched.

Lazy loading

The lazy loading pattern allows you to delay the creation of an object until you try to use it. This pattern can be implemented in any programming language. In Objective-C, we have used the property's getters and checked whether it was initialized. Swift adds support for lazy loading into the language, and this makes it even easier to apply this pattern. There are many things that can be lazy loaded, and we will cover them in this chapter.

Global constants and variables

Global variables and constants are always loaded lazily in Swift. This means that every global variable is initialized only when you access it for the first time. As a test, let's create a new `Person.swift` file and add this code to it:

```swift
struct Person {
  let name: String
  let age: Int

  init(name: String, age: Int) {
    self.name = name
    self.age = age
    print("\(name) Created")
  }
}

let Jon = Person(name: "Jon", age: 20)
let Sam = Person(name: "Sam", age: 28)
```

This file contains two global constants: `Jon` and `Sam`. We have also added a log statement to the `Person` structure's `init` method so that we can see when the person is created. Now let's try to access one of the global constants and see the console output:

```swift
print("Start!")
print("Age: \(Jon.age)")
//print("Age: \(Sam.age)")
```

```
Console Output:

Start!
Jon Created
Age: 20
```

As you can see, only the Jon global constant is created, and it's created when we try to access its age property. Yes, global constants and variables have powerful features in Swift, but you should almost never use global constants! There is a better way to do that.

Global variables are even more vulnerable because anyone can change them in the entire application.

Type properties

Both structures and classes can have a type property. The type property belongs to the type itself and not to an instance of that type. Only one copy of a type property will be created no matter how many instances of that type you create. They behave like global variables and constants, but they have a namespace scope of that type. Also, type properties can be declared as private and be hidden from rest of the application.

We can very easily improve our code by moving our global constants into the Person struct definition:

```
struct Person {
    let name: String
    let age: Int

    static let Jon = Person(name: "Jon", age: 20)
    static let Sam = Person(name: "Sam", age: 25)


    . . .
}
```

To access a type property, we need to prefix it with a type name that it is declared in; to access the Jon type constant, we would need to write Person.Jon. Classes have the same syntax and functionality for type properties as structures:

```
print("Age: \(Person.Jon.age)")
print("Age: \(Person.Sam.age)")
```

Lazy properties

There is also a way to lazy load an instance property. Let's extend our `Person` type and add a `HealthData` structure to it:

```
struct HealthData {
  init() {
    print("HealthData Created")
  }
}

struct Person {
  let name: String
  let age: Int

  var healthData = HealthData()
}
```

Now, every time we create a new person, it creates a `healthData` instance. In our example, `HealthData` could be a heavy object that connects to a database, fetches health data, and does a lot of work. And we don't need a `HealthData` structure to be created at the same time as `Person`; instead, we would like to create a `HealthData` structure only when we use it.

To make a property behave lazy, all you need to do is add the `lazy` attribute to its declaration:

```
lazy var healthData = HealthData()
```

The `lazy` instance property must be declared as a variable and not a constant.

It is very important not to abuse lazy instance properties. Accessing a lazy property adds a small performance overhead, for checking whether it was initialized or not.

That's why it is very important to analyze which parts of the system need to be created at the same time, because they are used at the same time, and which maybe not be used at all; so, we make them lazy to delay their creation.

When you access a lazy property, it actually mutates a value, so the instance has
to be declared as a variable. If you use a class with a lazy stored property, it can be
declared as a constant. This is because it's a constant reference and the changes are
applied to the value:

```
let ola = Person(name: "Ola", age: 27)
let health = ola.healthData // Error! It's mutating a value

var bobby = Person(name: "Bobby ", age: 5)
let bobbyHealth = bobby.healthData // Works fine

let someClass = SomeClass()
someClass.healthData // Works fine because class is a reference
type
```

An important thing to note about lazy properties is that they get initialized only
once. If you create a lazy optional property that you would like set to `nil` later and
initialize again, you will need to do it manually. Let's say `healthData` is a very
heavy instance and we want to clear it when it's not needed:

```
struct Person {
  ...
  lazy var healthData: HealthData? = HealthData()

  mutating func clearHealthData() {
    healthData = nil
  }
}

var ola = Person(name: "Ola", age: 27)
var health = ola.healthData //Get lazy loaded here
ola.clearHealthData()
health = ola.healthData // nil, nothing happens here.
```

As we said, the lazy stored property is initialized only once. So, the next time we
access it, after we have cleared it, there is no extra initialization code running. If
we really need this sort of lazy loading cache, then we will need to implement it
ourselves. It's not hard. We would need a stored property and a computed property
like this:

```
struct Person {
  ...
  private var _healthData: HealthData?

  mutating func clearHealthData() {
    _healthData = nil
```

```
    }

  var healthData: HealthData {
    mutating get {
      _healthData = _healthData ?? HealthData()
      return _healthData!
    }
  }
}
```

We have to create a mutating getter here, because the `var` properties can't be declared as `mutating`. Now, if we run our previous example once more, we will see that `HealthData` is created every time after cleaning it:

```
var ola = Person(name: "Ola", age: 27)
var health = ola.healthData //Get lazy loaded here
ola.clearHealthData()
health = ola.healthData // HealthData created again
```

Computed properties

The other way of delaying a property initialization is by using a computed property. As the name says, a computed property is computed every time you access it. It's important to remember that such a property will be computed every time, because this could have a negative impact on performance or if you perform any side effect on it.

The best use case of a computed property is when you want to provide a read-only property that uses internal data for its computation. A good example would be the full name of a person:

```
struct Person {
  let name: String
  let lastName: String
  let age: Int

  var fullName: String {
    print("calculating fullName")
    return "\(name) \(lastName)"
  }
}

var jack = Person(name: "Jack", lastName: "Samuel", age: 21)
print(jack.fullName)
print(jack.fullName)
```

It's really important to pay attention while performing any side effect or mutation operation on the computed properties. The mutating operation would require specifying explicitly for structures, but it would also be legal to do this for a class.

Lazy collections and evaluation

Another very interesting place where we could perform work lazily is collections and sequences. We store many elements in them, and sometimes, performing an operation such as `filter` or `map` would take a lot of time and may be unnecessary.

Before we dig into the details, let's first check out a small example to see why working lazily with a collection is so useful. We have a collection. We want to perform an operation such as mapping on it, and get one or a few elements from the result. Here is how we would implement it using an array:

```
let numbers = Array(1...1_000_000)
let doubledNumbers = numbers.map { $0 * 2 }
doubledNumbers.last
```

When we call the `map` method on the `numbers` array, it applies it to every element in the array and returns the new mapped array. As a result, we get a new `doubledNumbers` array. Our map `{ $0 * 2 }` closure is called as many times as there are elements in the array; in our case, it is 1,000,000 times. But we need only the last element from that array. Instead of mapping every element, we would like to map only the last one. For this situation, it is better to use a lazy collection:

```
let numbers = Array(1...1_000_000)
let lazyNumbers = numbers.lazy
let doubledNumbers = lazyNumbers.map { $0 * 2 }
doubledNumbers.last
```

The only difference here is that we added a `lazy` method call to create `LazyCollection` from an array:

```
public var lazy: LazyCollection<Self> { get }
```

When we call a map method on the lazy collection, it does not perform a mapping of every element immediately, but delays it and returns another `LazyCollection`— `LazyMapCollection`. The next time when we ask `doubledNumbers` for the last number, it performs a mapping to the last number and returns it to us. As a result, we call our map only once, which is exactly what we needed.

All `LazyCollections` and `LazySequences` work by the same principle; they do work on demand when you pull an element from it. When you call any method that is supposed to perform an operation on a sequence, such as `map` or `filter`, it is not performed immediately. Instead, the `LazySequence` saves the operation that needs to be performed and returns a new `LazySequence`. The `LazySequence` performs that operation only when you pull data out of it, like when you ask for the last element.

Sequences and collections

For a better understanding of laziness, it's useful to learn the difference between sequences and collections. We can apply lazy operations to both of them, but collections allow us to do more, and we will see why.

Sequences

A sequence is represented as a traversable set of elements. The main operation we would do on a sequence is iterating over its elements by starting from the first element and moving forward. For us to do this, the sequence uses a `GeneratorType` protocol. The `GeneratorType` protocol has a `next` method, which returns the next element, or nil if there is no element. This is the only method that is available on `GeneratorType`:

```
mutating func next() -> Self.Element?
```

The `SequenceType` and `GeneratorType` protocols are very simple. Each of them requires only one method to implement:

```
protocol SequenceType {
    func generate() -> Self.Generator
}
protocol GeneratorType {
    mutating func next() -> Self.Element?
}
```

Here is the simplest and most fundamental operation that is available for a SequenceType instance, an iteration over its elements:

```
let seq = AnySequence(1...10)
for i in seq {
  i
}
```

If you want your own types to be used in a for...in loop, you need to implement a SequenceType protocol for your type.

The other way in which we can iterate over the sequence is by manually getting the next element from the generator:

```
let gen = seq.generate()
while let num = gen.next() {
   num
}
```

The Swift standard library uses protocol extensions to provide extra functionality for the SequenceType. In Swift 2.0, SequenceType has become very powerful, with many methods, such as map and filter, sort, equal, and many others.

Collections

A collection, on the other hand, represents a group of items in which every item can be accessed by its index. The simplest example of a collection is an array. A collection type also implements a SequenceType protocol, because of which you can use a collection in every place where a sequence is expected.

CollectionType requires implementing three more methods in addition to the SequenceType protocol:

```
subscript (position: Self.Index) -> Self._Element { get }
var startIndex: Self.Index { get }
var endIndex: Self.Index { get }
```

CollectionType gives us random access to elements with a subscript method. Because the collection knows the first and last indexes, it can also calculate the size and iterate over its elements in both directions, from the beginning and from the end. This gives us much more power.

Implementing our own sequence and collection

As the wrap-up of getting to know sequences and collections, let's implement our own sequence and collection types. We can also add a print statement to it so that we can see in the console when it's actually doing work. This is very useful for inspecting the behavior of lazy collections. Let's start with the sequence:

```
struct InfiniteNums: SequenceType {

    func generate() -> AnyGenerator<Int> {
```

```
        var num = 0

        return anyGenerator {
            print("gen \(num)")
            return num++
        }
    }
}
```

Here is our simple sequence of infinite numbers. Our generator just returns the next integer. Now let's create a custom collection. It will be a bit more complicated than a sequence:

```
struct Collection10: CollectionType {
    let data = Array(1...10)

    var startIndex: Int {
        return data.startIndex
    }

    var endIndex: Int {
        return data.endIndex
    }

    subscript (position: Int) -> Int {
        print("Pos \(position)")
        return data[position]
    }

    func generate() -> AnyGenerator<Int> {
        var index = 0

        return anyGenerator {
            print("Col index: \(index)")
            let next: Int? = index < self.endIndex ? self.data[index++]
            : nil
            return next
        }
    }
}
```

For this example, we made an immutable collection of 10 integer elements. Its implementation is still quite simple. Because our collection is closed and it has the endIndex property, the generator needs to check the bounds, and it returns nil when there are no more elements in the sequence.

You can play around with those new types and use them with functions from the Swift standard library, which takes sequence or collection types as an argument. Now let's move on to the most interesting part; let's use these types with a lazy function.

Using lazy

Turning a collection or a sequence into a lazy version is very simple. Both sequences and collections have a `lazy` property, which returns a lazy version of that collection or sequence:

```
public var lazy: LazyCollection<Self> { get }
public var lazy: LazySequence<Self> { get }
[1, 2, 3].lazy
AnySequence(1...10).lazy
```

Using the lazy sequence

Let's now play with our infinite numbers sequence and the `LazySequence` map and filter methods:

```
let infNums = InfiniteNums()
let lazyNumbers = infNums.lazy

let oddNumbers = lazyNumbers.filter { $0 % 2 != 0 }
let doubled = lazyNumbers.map { $0 * 2 }
let mixed = lazyNumbers.filter { $0 % 4 != 0 }.map { $0 * 2 }

var gen = oddNumbers.generate()
var gen2 = mixed.generate()

for _ in 0...10 {
  gen.next()
  gen2.next()
}
```

The code here is pretty straightforward. We create a lazy sequence and apply a `map` and `filter` transformation to it. After that, we pull the first 10 results from it. Because the collection is lazy, it starts performing mapping and filtering only when we start pulling elements from it into the loop body, by calling the `next` method.

Because `LazySequence` is a `SequenceType` protocol, you can use it with any function that takes a `SequenceType` protocol or call any method that is available in that `SequenceType`. That operation won't be lazy and will return a result. If we try to use them with our infinity numbers, it could lead to an infinite loop. This is because our sequence is infinite:

```
lazyNumbers.contains(3) // returns true, stops when found
// lazyNumbers.minElement() // Infinite loop
```

Using a lazy collection

Working with lazy collections is very similar to working with lazy sequences. A `LazyCollection` implements a `CollectionType`, which means that we can use all the methods from `CollectionType` with a `LazyCollection`.

First, let's create a lazy collection and some additional helper functions to map and filter a collection that we will use:

```
let isOdd = { $0 % 2 != 0 }
let doubleElements = { $0 * 2 }

let col = Collection10()
let lazyCol = col.lazy
```

The `map`, `reverse`, and `filter` methods of a lazy collection have small differences. They return different lazy collection types: LazyMapCollection, LazyFilterCollection, and ReverseRandomAccessCollection:

```
lazyCol.map(doubleElements) //LazyMapCollection<Self.Elements, U>
lazyCol.reverse()
//LazyCollection<ReverseRandomAccessCollection<Self.Elements>>
lazyCol.filter(isOdd)   //LazyFilterCollection<Self.Elements>
```

Both the `LazyMapCollection` and `LazyFilterCollection` types implement a `LazyCollectionType`, and they have similar methods. Because of this, some lazy collections could have a different set of available methods.

One more interesting behavior of these lazy collections is the difference in the way some methods work. As an example, let's take a look at the `count` and `isEmpty` methods. First, let's try to use them with `LazyMapCollection`:

```
let lazyMap = lazyCol.map(doubleElements)
let count = lazyMap.count

lazyMap.isEmpty
lazyMap.reverse().isEmpty
```

Calling `count` and `isEmpty` does not force a lazy collection to perform any mapping operation. Because mapping doesn't change the number of elements of a source collection, it can use the underlying collections data — `startIndex` and `endIndex` — to compute a result for those methods.

However, calling `count` and `isEmpty` on a `LazyFilterCollection` does require the lazy collection to perform a filtering operation. For the `isEmpty` method, it stops as soon as it finds an element; but for the `count` method, it needs to apply the filtering operation to every element. In such cases, calling a `count` method would behave the same way as calling it on a regular collection:

```
lazyCol.filter(isOdd).isEmpty
lazyCol.filter(isOdd).count
```

One more difference between the `map` and `reverse` methods that is worth mentioning is in the subscript methods and indexes that they use:

```
//Query elements
lazyCol.map(doubleElements)[3]

let revCol = lazyCol.reverse()
let ind = revCol.startIndex.advancedBy(2)
revCol[ind]

let revMapCol = lazyCol.reverse().map(doubleElements)
let index = revMapCol.startIndex.advancedBy(2)
revMapCol[index]
```

`LazyMapCollection` allows us to use the original index that we used for the collection; in our case, this is an `Int` type. The `reverse` method returns a `ReverseRandomAccessCollection` with indexes of the `ReverseRandomAccessIndex` type. To use a subscript method on it, we need to use a `startIndex` or `endIndex` and an `advance` method to move the index to the needed position.

This was a general overview of lazy collections and sequences. However, there is one more very important characteristic of lazy types that we need to cover.

A lazy collection or sequence performs an operation, for example, a mapping of elements, when we start pulling elements from it. The difference here is that a regular array applies a mapping only once for every element and returns the result immediately.

A lazy collection applies a mapping every time we pull elements from it. If you are going to use a result of a mapping very often, then it may be better to use a regular array or add some kind of caching or memorization functionality. Let's see with the help of an example why it could be dangerous:

```
let col = Array(0...10)
let lazyCol = col.lazy

var x = 10
let mapped = lazyCol.map { $0 + x++ }

for i in mapped {
   print(" \(i)", terminator:"") //10 12 14 16 18 20 22 24 26 28 30
}

print("")
for i in mapped {
   print(" \(i)", terminator:"") //21 23 25 27 29 31 33 35 37 39 41
}
```

This example shows two important issues with using a lazy collection and a state in the map function. When we use a mapped lazy collection, we expect that the result will be the same for all loop iterations because we haven't changed anything. In reality, however, it will be different because the map function is called twice and the state we used in the map function will have been changed.

We could partially solve this problem by removing the state from the map function. Now the lazy mapped collection produces the same result, but even this doesn't change the fact that a map function will be executed in both loops. If we put an expensive operation inside the map function, it will make our app twice as slow:

```
// No state
let mapped = lazyCol.map { $0 + $0 + 10 }
```

Closing the topic of lazy collections, I would say that a lazy collection is very powerful, but you should used it as carefully as any other tool that helps improve performance.

Summary

In this chapter, we covered two more tools that can improve an application's performance: lazy loading and lazy execution techniques. First, we explained why it's important to make code behave lazily, when to do so, and in which way it can be implemented. Next, we showed you the lazy loading feature built into the Swift language and how to use it with global variables, type properties, and lazy stored properties.

In the rest of the chapter, we covered how to make a collection behave lazily. We used many functions with a lazy collection and showed the differences between a collection and a sequence.

In the next and final chapter, *Discovering All of the Underlying Swift Power*, we will take a look at some more advanced tools that will help you analyze the power of Swift, and go through a quick recap of what you learned throughout the book.

8
Discovering All the Underlying Swift Power

In previous chapters, we covered many topics about Swift, its powerful features, and how to improve an application's performance and make a solid application architecture. In this chapter, we will take a look at some tools and cover these topics:

- How Swift is so fast
- The Swift compiler and tools
- Analyzing the assembly code
- A recap of the important information learned
- Feature reading

How Swift is so fast

First, let's have a quick recap — that is, *why Swift is swift* — and see what is so special about Swift and its cool features.

Swift is a strongly typed compiled programming language. This makes it a very safe programming language. Swift is very strict about types, and it verifies that all types are used correctly in the source code. It catches a large number of issues at compile time.

Swift is also a static programming language. Its source code is compiled to the assembly code and the assembly code is compiled to the machine code using the LLVM tool. Running native machine code instructions is the fastest way of doing this. In comparison, Java and C# are compiled to a piece of intermediate code, and it needs a virtual machine to run it, or another tool that will translate it into machine instructions. Because Swift doesn't do this at runtime, it has a very big performance gain.

By mixing strongly typed rules and compiling to assembly code, Swift is able to analyze code very well and perform very good assembly code optimization. Swift is also built to be nice to write, with a pleasant and clean syntax and modern features and types. This unique combination of syntax, powerful features, safety, and performance makes Swift a very amazing programming language.

Swift command line tools

In this book, we have already worked with a terminal tool—a Swift REPL console. We start it by entering the xcrun swift command in the terminal. It starts the REPL, and we can enter Swift code and evaluate it.

xcrun

To start a REPL, we actually used two tools: xcrun and swift. xcrun is an Xcode command-line tool runner. It helps you run a command-line tool by its name from the active developer directory. When you have installed many versions of Xcode, you can select which one to use while executing a command-line tool. You can do this in Xcode by going to **Xcode | Preference | Locations | Command Line Tools**, or by running the xcode-select command from the terminal. In this way, xcrun allows you to avoid specifying the full path in the command-line tool you want to run, and makes the process of running it much simpler.

xcrun has a few more interesting options, and you should experiment with them. To get help, run the xcrun -h help command, which is usually available for most other command-line tools as well.

 Some command-line tools require the use of the full command notation: -help. An example of this is xcrun --help.

This means that by running the xcrun swift command, we are actually launching a swift command-line tool. If you run the xcrun swift -h command to get help, you will see that swift is actually a Swift compiler that has many options to choose from.

The other interesting feature that is available in xcrun is getting a path to the tool by its name. This is very useful because in this way, we can explore the folder in which this tool is located and find other available tools. As an example, let's see the swift command-line tool's location. To do this, you need to run the xcrun -f swift command, and the result is this:

```
/Applications/Xcode.app/Contents/Developer/Toolchains/XcodeDefault.
xctoolchain/usr/bin/swift
```

We can simply open this folder from the terminal by changing the directory with cd
and executing the open . command, which will open the current directory from the
terminal in the finder:

```
cd /Applications ... /usr/bin/

open .
```

In this folder, there are more than 50 different command-line tools available, such as
libtool, otool, swift, swift-demangle, and others. There is also another directory
that contains many interesting tools, such as ibtool, simctl, and xcrun:

```
/Applications/Xcode.app/Contents/Developer/usr/bin
```

We are not going to cover all of them here; we will leave it as homework for you to
play with them and explore their power.

Another way of discovering tools that are available for running in the current
directory through xcrun is by starting to type the commands and pressing the *Tab*
key. This will show suggestions from the available commands.

```
➜  bin  xcrun s
segedit*              strip*                swift-stdlib-tool*
size*                 swift*                swift-update*
strings*              swift-demangle*       swiftc@
```

The Swift compiler

In the previous step, while playing with xcrun, we discovered that there are two
different swift compiler tools: swift and swiftc. If you get help for with the -h
command, you will notice that they are both Swift compilers with similar options,
but there is a difference.

swift

The swift tool compiles and executes swift code. If you run it without any
arguments, it will launch a REPL and give you the ability to evaluate the Swift code.

```
➜  ~  xcrun swift
Welcome to Apple Swift version 2.0 (700.0.57 700.0.72). Type :help for assistance.
  1> 1 + 10
$R0: Int = 11
  2> print($R0)
11
  3>
```

You can also pass a Swift file that you want to run as a parameter. We can simply create a swift file from the command line:

```
echo 'print("Hello World")' > main.swift
xcrun swift main.swift
Hello World
```

You can also pass additional options, such as -O to enable optimization:

```
xcrun swift -O main.swift
```

swiftc

The `swiftc` compiler compiles swift code and produces the result, but it doesn't execute it. Depending on the option, you can get a binary file, an assembly, an LLVM IR representation, or something else.

Unlike the swift tool, `swiftc` has a one required parameter: an input swift file. If you try to run it without this parameter, you will get an error:

```
xcrun swiftc
<unknown>:0: error: no input files
```

If you run it and pass a Swift file as a parameter, it will produce an executable file. Let's use the same `main.swift` file:

```
xcrun swiftc main.swift
```

As a result, you get an executable file with the same name—main. If you run it, you will get the same output, `Hello World`, in the console. The interesting thing about `swiftc` is that it allows you to pass many Swift files as input, compile them together, and produce the executable. Let's create one more swift file, called `file.swift`, and add this function to it:

```
func bye() {
  print("Bye bye")
}
```

Now, we will edit our `main.swift` file and the call to the `bye()` function in it. If we try to compile the `main.swift` file now, it will show us an error:

```
xcrun swiftc main.swift
main.swift:2:1: error: use of unresolved identifier 'bye'
bye()
^
```

Not surprisingly, the Swift compiler can't find the `bye` function declaration and so it fails. What we need to do is compile both the `file.swift` and `main.swift` files at the same time:

```
xcrun swiftc main.swift file.swift
```

The order of the files that you pass to `swiftc` is not important; we could have called them as `swiftc file.swift main.swift` as well. If you run this executable file now, you will see two sentences in the console:

```
Hello World
Bye bye

[Process completed]
```

Now that you know how to use a Swift compiler, let's move on to the fun part. Let's use a swift compiler to produce different results. For simplicity, we will merge our swift code with the `main.swift` file and add more instructions to get a more interesting result. Here is the final code:

```
func bye() {
  print("bye")
}

print("Hello World")

let a = 10
let b = 20
let c = a + b
print(c)
bye()
```

The Swift compilation process and swiftc

The compilation of the Swift source code is quite an interesting process, and it involves several steps. The Swift compiler uses LLVM for optimization and binary generation. To better understand the entire process, refer to this flow diagram:

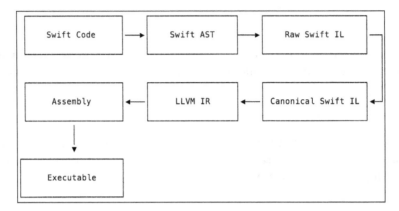

First, the Swift source code is transformed into an **AST** (short for **Abstract Syntax Tree**). Then, it is transformed into **SIL** (short for **Swift Intermediate Language**), first into a raw SIL and then into a canonical SIL. After that, it is transformed into LLVM **IR** (short for **Intermediate Representation**). In this step, LLVM takes care of the rest. It takes IR, does an optimization, and produces an assembly and, after that, an executable for a specific architecture.

The interesting part in preceding diagram is the steps for generating SIL. It's a Swift-specific optimization and it was created specifically for swift. Other programming languages, such as C, don't do this optimization before they generate LLVM IR, and they have one less optimization step.

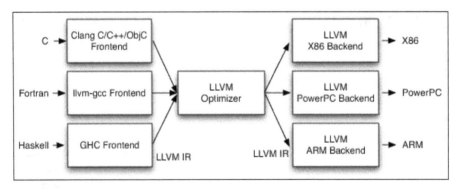

With `swiftc`, it's possible to generate the results for each of those steps. It's incredibly useful for code optimization analysis. To see all the available modes, just run `xcrun swiftc -h`. Now, let's quickly take a look on them.

Swift AST

Swiftc has three different options for generating AST. Each of them generates AST with different levels of details. The AST code representation shows us how the Swift compiler sees and analyzes the code:

```
xcrun swiftc -dump-ast main.swift
xcrun swiftc -dump-parse main.swift
xcrun swiftc -print-ast main.swift
```

The output of `-dump-ast` contains the maximum details, and it could be hard to analyze. Let's take a look at the `-dump-parse` example first:

```
(source_file
  (var_decl "a" type='<null type>' let storage_kind=stored)
  (top_level_code_decl
    (brace_stmt
      (pattern_binding_decl
        (pattern_named 'a')
        (integer_literal_expr type='<null>' value=10)))
```

This AST code represents the `var a = 10` Swift code. Each instruction is parsed into a separate tree node and then put together in a tree representation. You can find more information about Clang's AST at `http://clang.llvm.org/docs/IntroductionToTheClangAST.html`.

SIL

The **Swift Intermediate Language** (**SIL**) is one of the most useful tools for analyzing Swift code. It contains many details and is very readable and easy to analyze. For generating a SIL, xcrun has two modes; `-emit-silgen` generates raw SIL and `-emit-sil` generates canonical SIL:

```
xcrun swiftc -emit-silgen main.swift
xcrun swiftc -emit-sil main.swift
```

Raw SIL and canonical SIL are almost the same. Raw SIL is a bit simpler and it doesn't include the details of private function implementations and some global objects. Let's take a look at the generated raw SIL:

```
sil_stage raw

import Builtin
import Swift
import SwiftShims

// main.a : Swift.Int
sil_global [let] @_Tv4main1aSi : $Int

...

// main
sil @main : $@convention(c) (Int32, UnsafeMutablePointer<UnsafeMutable
Pointer<Int8>>) -> Int32 {
...
}

// main.bye () -> ()
sil hidden @_TF4main3byeFT_T_ : $@convention(thin) () -> () {
...
}

// static Swift.+ infix (Swift.Int, Swift.Int) -> Swift.Int
sil [transparent] [fragile] @_TZFSsoi1pFTSiSi_Si :
$@convention(thin) (Int, Int) -> Int
```

A really nice feature of SIL is that it contains comments that explain the generated code. The `let a: Int` statement would be translated into `@_Tv4main1aSi : $Int` and we can see this from the comment that stays above the generated SIL:

```
// main.a : Swift.Int
sil_global @_Tv4main1aSi : $Int
```

The SIL represents Swift code in a mangled format. The names contain a lot of information about the type, the count of symbols in the name, and so on. Some mangled names can be very long and really hard to read, such as `_TZvOSs7Process11_unsafeArgvGVSs20UnsafeMutablePointerGS0_VSs4Int8__`.

We can demangle a name back to its normal notation with the `swift-demangle` tool. Let's try to demangle `@_Tv4main1aSi` and see whether it really translates into `main.a : Swift.Int`:

```
xcrun swift-demangle _Tv4main1aSi
_Tv4main1aSi ---> main.a : Swift.Int
```

If you want to learn more about name mangling, you can read a great post about it written by Mike Ash at `https://mikeash.com/pyblog/friday-qa-2014-08-15-swift-name-mangling.html`.

LLVM IR

Intermediate Representation (**IR**) is a more low-level code representation. It is not as human-friendly and readable as SIL. This is because it has more information for the compiler than for humans. We can use IR to compare different programming languages. To get Swift's IR, use the `-emit-ir` options, and to get IR for C, we can use `clang -emit-llvm`:

```
xcrun swiftc -emit-ir main.swift
clang -S -emit-llvm  main.c -o C-IR.txt
```

Other swiftc options

The `swiftc` compiler is very powerful and has many more modes and options. You can create an assembly, a binary, a linked library, and object files. You can also specify many options, such as an output file with the `-o` option, optimization `-O`, `-Onone`, and many others:

```
xcrun swiftc -emit-assembly main.swift -o assembly
```

Analyzing executable files

It is very difficult to analyze the assembly code generated by the swiftc compiler. To make our lives easier, we will use a Hopper Disassembler tool to disassemble executable files, generating a piece of pseudocode and analyzing it. You can download the free version of Hopper from `http://www.hopperapp.com`.

The Hopper Disassembler tool can work with binary, executable, and object files. The easiest way of using it is by generating an executable file with the `swiftc main.swift` command and opening it in Hopper. You can simply drag and drop the `main` executable file to open it in Hopper.

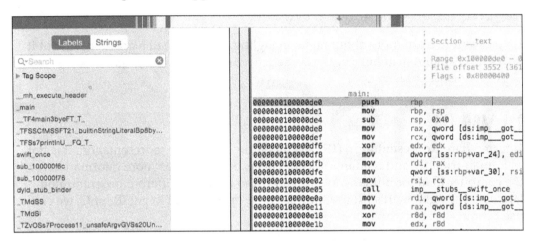

On the left-hand side, you can find all the labels for functions and variables and navigate to them. The search feature is very useful when you are analyzing a big project with many functions. In the center is an assembly code; you can press *Alt + Enter* to see the pseudocode for the current procedure. It is much easier to analyze high-level pseudocode.

We can also compile an application in Xcode and disassemble our `SimpleApp.app` in Hopper. This allows us to analyze very large and complex applications as well.

As an experiment, let's compile the same swift file in two ways—with optimization enabled and without it—and compare the generated assembly code. Thus, you will see the power of the optimization option:

```
swiftc main.swift -O -o mainOptimized
```

Summary

This chapter illustrated the Swift compiler, command-line tools, and the process of compilation of the Swift source code. Knowing the available tools and mastering them is very important because it makes you much more productive. Xcode has many tools, and we showed you how to find and use them.

For analyzing and optimizing Swift code, it's very useful to know and understand the compilation process. In this chapter, we took you through an entire journey of the Swift compiler, starting from the source code and ending with the executable file. We also showed you how to get the result for a specific complication step, such as getting the SIL or IR code representation.

Final thoughts

Our learning journey in this book comes to an end, and now you have mastered Swift's techniques for creating high-performance applications. Let's have a quick recap.

By now, you have learned how to use the power of Swift and optimize your Swift code, but you should remember the main rule of optimization – optimize only when needed, not upfront.

Solid architecture and well-structured and clean code are two of the most important characteristics of a good application. We have been expressing this throughout the book, and an entire chapter (*Chapter 2, Making a Good Application Architecture in Swift*) was dedicated to it.

Performance optimization doesn't always have to bring much complexity to the source code. Sometimes, applying small changes, such as adding the `@noescape` attribute, removing a few `print` statements, using the correct data structures, and other techniques presented in the book, could improve performance with no negative effect on the source code. Sometimes, the source code can become even cleaner and more readable; for example, by using `@noescape`, we don't need to explicitly specify `self.` when referencing instance members.

Learning Swift's features and tools that are at your disposal before you start creating an application is very important. It is much easier to create a good application from the start by spending some time on planning and preparation than to try to refactor it and fix performance and code architecture issues later.

Now, you are ready for the journey of creating incredible applications with Swift!

Index

Symbols

@autoclosure attribute 148-151
@noescape attributes 148
@noescape keyword 149, 150
@testability
   enabling 73

A

Abstract Syntax Tree (AST) 176
Accelerate
   about 122
   URL 122
application architecture
   achieving 140
   component with single responsibility,
      identifying 141
   comporelations and communication,
      building 141-147
   concurrent code execution 141
application development
   phases 59
arrays
   about 112
   elements, accessing 113
   elements, appending 113
   fast operations 113
   search operation 114
   slower operations 113
   sort operation 114
   using 113
asynchronous code
   designing 134-136
Automatic Reference Counting (ARC) 99

C

Clang's AST
   URL 177
clean code, Swift features
   clean closure syntax 6
   default and memberwise initializers 7
   default parameter values 6
   external names 6
   guard 8
   no semicolons 5
   operators 8
   subscripts 7
   type inference 5, 6
closed range operator 14
closure expression
   about 45
   closure syntax, trailing 46
   implicit return type 46
   shorthand argument names 46
   type inference 45
code, characteristics
   code performance 24
   code quality 24
   performance 25
collections
   about 124, 162, 163
   implementing 163, 164
collections memory allocation
   about 119
   default values 119, 120
   empty 119
   reserve capacity 119

value types
  about 35
  state, representing with classes 36, 37
  structures 35, 36
variables
  about 80, 81
  differentiating, with constants 28-30

X

Xcode
  REPL 68-70
XCPlayground
  about 67
  XCPCaptureValue 67
  XCPSetExecutionShouldContinue
     Indefinitely 67
  XCPSharedDataDirectoryPath 67
  XCPShowView 67
xcrun 172, 173

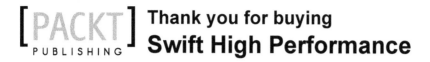

**Thank you for buying
Swift High Performance**

About Packt Publishing

Packt, pronounced 'packed', published its first book, *Mastering phpMyAdmin for Effective MySQL Management*, in April 2004, and subsequently continued to specialize in publishing highly focused books on specific technologies and solutions.

Our books and publications share the experiences of your fellow IT professionals in adapting and customizing today's systems, applications, and frameworks. Our solution-based books give you the knowledge and power to customize the software and technologies you're using to get the job done. Packt books are more specific and less general than the IT books you have seen in the past. Our unique business model allows us to bring you more focused information, giving you more of what you need to know, and less of what you don't.

Packt is a modern yet unique publishing company that focuses on producing quality, cutting-edge books for communities of developers, administrators, and newbies alike. For more information, please visit our website at www.packtpub.com.

Writing for Packt

We welcome all inquiries from people who are interested in authoring. Book proposals should be sent to author@packtpub.com. If your book idea is still at an early stage and you would like to discuss it first before writing a formal book proposal, then please contact us; one of our commissioning editors will get in touch with you.

We're not just looking for published authors; if you have strong technical skills but no writing experience, our experienced editors can help you develop a writing career, or simply get some additional reward for your expertise.

Swift Essentials

ISBN: 978-1-78439-670-1 Paperback: 228 pages

Get up and running lightning fast with this practical guide to building applications with Swift

1. Rapidly learn how to program Apple's newest programming language, Swift, from the basics through to working applications.

2. Create graphical iOS applications using Xcode and storyboard.

3. Build a network client for GitHub repositories, with full source code on GitHub.

Learning iOS 8 Game Development Using Swift

ISBN: 978-1-78439-355-7 Paperback: 366 pages

Create robust and spectacular 2D and 3D games from scratch using Swift – Apple's latest and easy-to-learn programming language

1. Create engaging games from the ground up using SpriteKit and SceneKit.

2. Boost your game's visual performance using Metal - Apple's new graphics library.

3. A step-by-step approach to exploring the world of game development using Swift.

Please check **www.PacktPub.com** for information on our titles

www.ingramcontent.com/pod-product-compliance
Lightning Source LLC
Chambersburg PA
CBHW060558060326
40690CB00017B/3749

9 781785 282201